Straight Talk About Stress

A Guide for Emergency Responders

Mike McEvoy

EMS Coordinator
Saratoga County, New York

NFPA®

National Fire Protection Association
Quincy, Massachusetts

Product Manager: Pam Powell
Developmental Editor: Robine Andrau
Editorial-Production Services: Chestnut Hill Enterprises, Inc.
Interior Design: Cheryl Langway
Composition: The Format Group, LLC
Photo Research: Josiane Domenici
Cover Design: Cameron, Inc.
Manufacturing Manager: Ellen Glisker
Printer: Courier/Westford

NFPA No.: STRESS04
ISBN: 0-87765-481-6
Library of Congress Card Catalog No.: 2003111408

Printed in the United States of America
04 05 06 07 08 5 4 3 2 1

To Battalion Chief Thad Dahl of Vashon Island Fire & Rescue in Washington State, whose spirited battle with esophageal cancer taught me countless lessons about the inescapable price we pay for stress in our lives

Contents

Preface   ix

Chapter 1
Different Faces of Stress   1

    Stress Defined   *1*
    Physical and Psychological Stress   *2*
    Stress from Positive and Negative Events   *9*
    Different Responses to the Same Event   *14*
    Summary   *16*
    References   *17*

Chapter 2
Traits of Emergency Responders   19

    Role of Adrenaline   *19*
    Traits Common to Emergency Responders   *21*
    Negative Aspects of Emergency Responder Traits   *30*
    Summary   *32*
    References   *32*

Chapter 3
Stress on the Job   33

    Occupational Stress Versus Job Stress   *33*
    Symptoms of Overstress   *34*
    Control Orientation as Source of Conflict   *35*
    Conflict Management   *36*
    High Expectations as Cause of Stress   *41*
    Strengths of Emergency Responders   *44*

Summary   *44*
References   *45*

Chapter 4
Critical Incident Stress   47

Acute Stress Disorder (ASD) and Post-Traumatic
    Stress Disorder (PTSD)   *48*
Effects of Critical Incidents   *49*
Categories of ASD/PTSD Symptoms   *51*
Responses to ASD/PTSD Symptoms   *52*
Critical Incident Stress Debriefing (CISD)   *53*
Summary   *61*
References   *62*

Chapter 5
Stress at Home   63

Understanding the Source of Home Life Problems   *63*
Issues Affecting Home Life   *65*
Communication as Key Factor   *69*
Addressing Behavior Problems   *77*
Summary   *78*
References   *79*

Chapter 6
Shift Work and Sleep   81

Shift Work and Sleep Deprivation   *81*
Sleep-Wake Cycle   *84*
Sleep Hygiene   *89*
Summary   *93*
References   *94*

Chapter 7
**Development of Competence in Emergency
Responders   97**

Dreyfus Model of Skill Acquisition   *97*
Performance Under Pressure   *98*
Stages of Competence Development   *105*

Development of Competence   *117*
Summary   *119*
References   *120*

Chapter 8
Personal Stress Management Program   121

Overview of Stress Management   *121*
The Body and Stress   *124*
The Mind and Stress   *129*
Interaction with Others and Stress   *135*
Summary   *139*
References   *140*

Index   143

Preface

Fire fighters, emergency medical services (EMS) providers, and law enforcement officers deal with stress every day. Those who contemplate joining or who are just starting work in the emergency services sector will become proficient at helping others in their most stressful moments. Rarely will the people that emergency responders meet in calls for service be having a good day. Emergency responders come to expect seeing the public in trouble and routinely apply their knowledge and skills to help others manage crisis.

Other occupations such as surgery, anesthesiology, critical care nursing, and airline traffic control share characteristics with the emergency services sector. Anesthesiology, for example, is often described as 90 percent sheer boredom and 10 percent sheer terror—an adage that could apply equally well to fire fighting, EMS work, and law enforcement. Although comparable, other occupations are simply not the same as emergency services. They may attract similar personalities and have related demands, but one critical element separates emergency services from all other careers: the bond of brotherhood/sisterhood. None of us has probably ever seen (and likely never will see) thousands of anesthesiologists, ICU nurses, or airline traffic controllers lining the streets of a community to pay their last respects to a fallen coworker. It simply doesn't happen. Yet such a funeral is familiar in the emergency services sector.

People in some occupations risk being exposed to extreme stress. Fire fighting, EMS work, and law enforcement are three such occupations. Because emergency responders help people in stressful situations on a daily basis, we would expect them to have the skills to deal with any level of stress in their own lives. This might be true in some cases; more often, however, the personality characteristics that make a good fire fighter, medic, or law enforcement officer also lead to increased stress for that

individual. Frequently, those drawn to so-called high-stress occupations have higher incidences of divorce and family problems. They are also more likely to die from stress-related diseases and are prone to overeating and chemical recreation
using alcohol or drugs.

Straight Talk About Stress is not a typical how-to manual
for managing stress in one's life or in the lives of others. There
is no "psychobabble" or complicated terminology. Cure-alls such
as diet, exercise, relaxation techniques, and time management
touted by stress management gurus are not discussed in detail
in this book. These techniques, as well as yoga, meditation, and
biofeedback information, can be found in many other books
about stress. Instead, *Straight Talk* is about what makes the
fire fighter, the medic or EMS provider, and the law enforcement officer tick and what probably drew him or her into emergency services in the first place. Certain types of people are attracted to fire, EMS, and police work; others are not and, in all
likelihood, would not be able to do the job no matter how hard
they tried. Emergency responders are different from others,
and these differences give them the tools they need to succeed
in emergency services careers. These same differences can also
easily lead to conflict, communication problems, and frequent
disappointment. *Straight Talk About Stress* explains how these
problems might happen and what emergency responders can do
to control them.

Chapter 1, "Different Faces of Stress," defines stress and discusses how different people respond differently to the same
event. Chapter 2, "Traits of Emergency Responders," sketches
out the unique characteristics that emergency responders share
and describes the role of adrenaline. Chapter 3, "Stress on the
Job," differentiates occupational stress and job stress and offers
positive measures for managing conflict. Chapter 4, "Critical
Incident Stress," outlines categories of and responses to acute
stress disorder (ASD) and post-traumatic stress disorder (PTSD).

Chapter 5, "Stress at Home," looks at the stressors in the
home life of emergency responders. Chapter 6, "Shift Work and
Sleep," examines the effects of sleep deprivation on responders
and some sleep hygiene measures they can use to counteract
those effects. Chapter 7, "Development of Competence in
Emergency Responders," looks at the stress that accompanies
emergency responders as they progress through the stages of

competence from the novice to the proficient responder—and for some to the expert level. Finally, Chapter 8, "Personal Stress Management Program," presents an overview of stress management. It discusses the relationship of stress and the body, the mind, and the interaction with others and provides guidance on developing one's personal stress management program.

This book is further enhanced by photos and drawings and by two features— "Key Point" margin notes, which highlight important information, and "On the Lookout" margin notes, which suggest actions the emergency responder can take to counter stress.

We have all undoubtedly met some very happy people during our lifetimes. Everyone probably knows individuals who always seem happy, no matter what happens around them. We may even wonder how such a person can always be in a good mood. The answer is simple: Happy people choose to be happy. Everyone probably also knows some miserable individuals. The same answer applies to miserable people: They choose to be miserable. Most of us fall somewhere in between being happy all the time and being miserable all the time. Many of us would like to be happier. *Straight Talk* will provide the tools emergency responders need to choose happier, healthier, and less stressful lives. Fire fighters combating a huge blaze are in their glory, medics thoroughly enjoy working a major trauma, and cops find a high-speed pursuit exhilarating. These events don't happen every day, however, and for some may occur only once in a career. *Straight Talk* is meant to help emergency responders make the most of their careers during the times in between those high points.

Friendships, family, and relationships also bring happiness into our lives. The relationships of emergency responders are often affected by a career in emergency services, many times adversely so. Unfortunately, when a relationship gets rocky, there is added stress. Stress destroys relationships and leads to loneliness. This book explains how to keep relationships alive and well. Knowing as much as possible about emergency responders, their personality, and how they interact with others improves awareness. The goal of *Straight Talk About Stress* is to give emergency responders greater control over their jobs, their interactions and communications with others, and their relationships.

ACKNOWLEDGMENTS

I owe special thanks to the countless fire, law enforcement, and medical professionals worldwide who have shared their stories, struggles, and triumphs. Your strength in the face of adversity inspires others. I am proud to call you my family.

ABOUT THE AUTHOR

 Mike McEvoy is EMS Coordinator for Saratoga County, New York, a paramedic for Clifton Park-Halfmoon Ambulance Corps, a fire fighter, and the medical advisor for West Crescent Fire Department. A former forensic psychologist with the U.S. Department of Justice, Mike has worked since 1988 as a critical care nurse in the Cardiac Surgery Intensive Care Units at Albany Medical Center while teaching pulmonary and critical care medicine at Albany Medical College in upstate New York. He is a member and officer of the New York State EMS Council and New York State's Emergency Medical Advisory Committee. In 2002, Mike became the first EMS chief elected to the board of directors of the New York State Association of Fire Chiefs.

Mike's interest in stress began while he was working as a medic in the Lower East Side of Manhattan during the late 1970s. This interest has driven him to develop critical incident stress management programs for emergency responders, hospital and medical staff, transportation, and industry. Mike is a frequently published author in fire, EMS, nursing, and medical journals and speaks worldwide on leadership, stress management, medicine, and new technologies. In his free time, he is an avid hiker and winter mountain climber. You can contact Mike at mcevoymike@aol.com.

Different Faces of Stress

1

S tress is hard to pin down. Too much of it can kill you; yet no person can survive without it. Athletes and other performers and competitors maximize their performance by keeping stress at an optimal level. Too little or too much anxiety and stress causes a drop in efficiency [1]. People with no stress are buried in cemeteries all over the world. Stress for all of them ended when they died, and for some, it was stress itself that killed them. In order to battle stress, you need to know what it is and where to find it.

STRESS DEFINED

Selye's Definition of Stress

Hans Selye, a Canadian physician and researcher, studied stress for over forty years. His contributions to science and medicine have led many to refer to him as the "Father of Stress." Selye also thought that stress was difficult to understand, but he eventually arrived at a scientific definition for stress. He defined stress as the response the body makes to a demand for change [2]. His definition remains popular today.

Stress is more than simply a vocabulary word for members of the emergency services; it's a lifestyle and a challenge that comes with the job. The effects of stress are all around us, and we see these effects in new and different ways every day. Helping others cope with stress is a significant component of emergency services work.

Many people think of stress as something bad—a negative that we could all do without. Not all stress, however, is

> **Key Point:** *Stress is the response the body makes to a demand for change.*

bad, or even problematic. In fact, stress brings excitement and challenges and helps us enjoy living. In order to live and work with stress, we need a basic understanding of what it is we're up against. What demands are placed on our bodies to change and how do we typically respond to those demands?

Stress as a Reaction

Researchers continue to struggle today in an effort to agree on exactly what stress is and how it can be measured. One thing we do know is that stress is not a person or an event but rather our reaction to that person or event. A spouse, boss, child, dog, and car are not stressful by themselves. Stress is how we react to the people, things, and events in our lives. That explains why different people experience different levels of stress from the same event: Every one of us reacts differently.

> **Key Point:** *Stress is not a person or an event but rather our reaction to that person or event.*

PHYSICAL AND PSYCHOLOGICAL STRESS

Physical Stress

Prehistoric people probably noticed that certain physical changes in their environment produced corresponding changes in their bodies. When they were cold, they shivered. When they ran great distances or otherwise exerted themselves, no doubt their hearts pounded and they perspired just as we do today. These physiological responses formed the backbone of what was first described by Hans Selye in 1936 as the general adaptation syndrome (GAS).

General Adaptation Syndrome (GAS)

Exposure to a stressor produces three distinct phases in the general adaptation syndrome: alarm, resistance, and exhaustion (see Figure 1.1). In the first, or alarm, phase, our body kicks into high gear in an attempt to adapt or survive the stressor. Breathing and heart rate increase, our muscles tense,

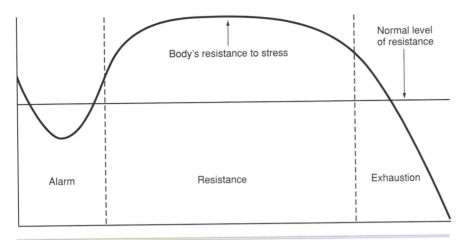

Figure 1.1 *Hans Selye's General Adaptation Syndrome*

pupils dilate, alertness heightens, and our body diverts blood to the skeletal muscles and brain. During this time, our overall resistance is somewhat weaker, which makes our system increasingly vulnerable to other stressors should any of them attack us during this phase.

During the second, or resistance, phase our body has adapted to the stressor and spends a significant amount of energy resisting its effects. The physical signs and symptoms of the alarm phase are gone, and, overall, resistance to the stressor is at a very high level. Resistance to other stressors, however, is diminished, and the body continues in a state vulnerable to attack by another stressor.

The third, or exhaustion, phase results from long-term exposure to a specific stressor that has left our body in a continued state of resistance (the second phase). Eventually, the energy needed to remain in the resistance phase runs out, leading to exhaustion. In this phase, some symptoms of the alarm phase may recur, and our body may succumb.

A fire fighter operating at the scene of a working restaurant fire provides a short-term example of the GAS. Responding to the call, the fire fighter experiences a heightened sense of energy (alarm) listening to radio reports of flames ripping from the rear of the building. His excitement continues as he dons his self-contained breathing apparatus and enters the structure on a hose line. Hours later, during overhaul, our fire fighter has adapted to the scene and no longer finds himself experiencing a high level of excitement (resistance). Nearly half a day later, the fire fighter finally returns to his station and begins the arduous task of washing hose and cleaning the apparatus. At this point, he finds himself drained of energy (exhaustion) and nearly too tired to continue working.

A paramedic working in a busy urban EMS system provides a long-term example of the GAS. In her first year on the street, our paramedic is excited and energized by the high call volume, violent trauma, and the occasional danger she encounters (alarm). During her second and third year, she is no longer as enthusiastic about her work and at times spends great effort trying to maintain a positive outlook on the job (resistance). In her fourth year, she finds herself increasingly frustrated or "burnt out" at work and believes that she is losing the energy needed to continue (exhaustion).

Hans Selye used the general adaptation syndrome to illustrate that continued stress over time leads to illness, which leads to disease, which leads to death (see Figure 1.2).

Fight or Flight

Today, we have many more clues as to how the body responds to stress. The process is similar whether the stress is short term or long term. For example, if a large bear were to suddenly burst into the room you were sitting in, your body would immediately enter a stage of alarm. Physical and biochemical changes such as increased heart and respiratory rates would lead to greater alertness and the ability for you to fight, freeze, or run, as well as to be sufficiently afraid.

This constellation of fright, fight, freeze, and flee are modern day simplifications of the first phase of the GAS (i.e., alarm), which was described physiologically in 1939 by American physiologist Walter Cannon as "fight or flight." [3] Energy used to respond to a threat continues to be used as you move into the second phase of the GAS (i.e., resistance). In our example, if the bear were to continue threatening you for a pro-

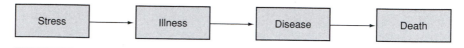

Figure 1.2 *Outcome of Continued Stress*

longed period of time, you would very likely become exhausted and enter the third phase of the GAS. As in this example, you can readily see how a hunted animal eventually weakens and can no longer escape its hunters. You might also be able to apply the GAS paradigm to prolonged work at the scene of a major mass casualty incident, when rescuers are unable or unwilling to rest (see Figure 1.3).

Chronic Stress

Smaller incidents and stress of a more chronic nature follows the GAS pattern as well. More people die from chronic stress than perhaps any other medical or psychological malady [4]. The way people die from chronic stress is subtler and can take many years to occur. The complex process by which chronic stress affects the body involves the nervous and endocrine systems interacting to release several hormones that increase energy and strengthen the body's defenses. When individuals experience long-term or chronic stress, their bodies enter a prolonged phase of resis-

> **Key Point:** *More people die from chronic stress than perhaps any other medical or psychological malady.*

Figure 1.3 *Fire Fighters at American Airlines Flight 587 Crash in Queens, New York, 2001 (Source: AP Photo/William C. Lopez)*

tance during which these hormones are continually released. As early as the 1800s, physicians noted that the body's response to the stress of severe burns often led to ulcers [5].

Role of Stress in Illness and Disease

Today, we are increasingly enlightened about the role of stress in causing illness and disease. Activation of the sympathetic nervous system interferes with metabolism of fats in the body and can increase cholesterol levels, leading to greater incidence of heart disease and stroke. The ability of the body's immune system to fight infection is suppressed with both chronic and acute stress, increasing the likelihood of common colds and flu as well as aggravating other immune system–related disorders such as allergies, arthritis, and perhaps even some cancers [4].

High blood pressure is often related to continual stress-related stimulation of the nervous system over time. High blood pressure causes circulation problems and heart damage and injures the kidneys. Diabetes, asthma, certain skin rashes, and irritable bowel syndrome can all be precipitated or worsened by chronic stress. Migraine and tension headaches, temporomandibular joint dysfunction (TMJ), and irritable bowel syndrome (IBS) are all considered to be stress-related problems (see Table 1.1). The simplest explanation takes us right back to the cemetery. Many of the people buried there died from stress-related diseases. There can be no doubt: Too much stress can kill you [2, 4].

Table 1.1   *Medical Conditions and Diseases Linked to Stress*

Nervous System Overstimulated	*Immune System Weakened*
Headaches (tension, migraine)	Common colds and the flu
Stomach ulcers and colitis	Allergies
Irritable bowel syndrome (IBS)	AIDS
High blood pressure	Cancers
Heart disease	Systemic lupus (SLE)
Asthma	Arthritis
Temporomandibular joint dysfunction (TMJ)	

Source: Based on Brian Luke Seaward, *Managing Stress: Principles and Strategies for Health and Wellbeing,* 3rd ed., Sudbury, MA: Jones and Bartlett Publishers, 2002.

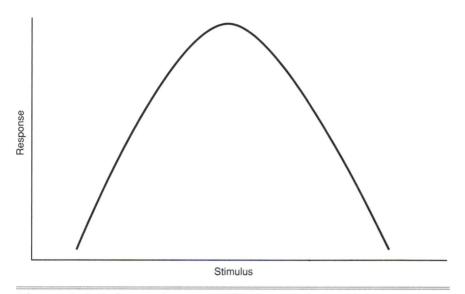

Figure 1.4 *Yerkes–Dodson Curve Illustrating Relationship Between Stress and Performance*

Psychological Stress

Stress is not only physical, but psychological as well. A certain level of stress increases alertness, problem-solving ability, creativity, and motivation. The Yerkes-Dodson curve is commonly used to illustrate the relationship between stress and performance (see Figure 1.4). Table 1.2 shows the psychological effects of low-level stress and of too much stress. Low-level stress can improve gross motor skills, decrease reaction time, and enhance self-confidence.

Table 1.2   *Psychological Effects of Stress*

Effects of Low-Level Stress	*Effects of Too Much Stress*
Increased alertness	Tunnel vision
Increased problem-solving ability	Increased symptoms of anxiety
Increased creativity	Decreased tolerance for pain
Increased motivation	and frustration
Improved gross motor skills	Inability to think clearly
Decreased reaction time	Increased mistakes leading to
Enhanced self-confidence	injuries

Accelerated visual processing that produces a "slow-motion effect" is believed related to small amounts of stress. Too little stress leaves people tired and bored. Too much stress causes "tunnel vision," a common emergency services term that refers to focusing on small details at the expense of missing a bigger, more important picture. A classic example is focusing on an injured child inside an overturned automobile—and missing the live electrical wires touching the outside of the car. Too much stress also causes increased sensitivity to symptoms of anxiety, decreased tolerance for pain and frustration, loss of ability to think clearly, and increased mistakes that often lead to injuries. You need only to look at the faces of people fleeing a burning or collapsing building to understand that stress has a strong psychological as well a physical component (see Figure 1.5).

Figure 1.5 *People Fleeing from Terrorist Attack on New York's World Trade Center (Source: AP Photo/Suzanne Plunket)*

STRESS FROM POSITIVE AND NEGATIVE EVENTS

We typically associate stress with negative events, but positive events can be stressful as well. One of the earliest and best illustrations of the comparable stress value of life events was a ranking survey done in 1967 by two psychiatrists—Thomas Holmes and Richard Rahe—from the University of Washington Medical School in Seattle.

Holmes and Rahe Stress Survey

Holmes and Rahe asked 394 people to rank 43 life events according to the intensity and length of time necessary to recover from each event. The results were later assembled in a test to predict the likelihood of illness. Although the tool is useful as a predictor of illness, it may be even more valuable as a comparison of life events and their perceived stress values.

Table 1.3 is an adaptation from the Holmes and Rahe stress survey [6]. In the left column are life events that are normally considered happy occurrences, such as marriage, buying a new home, retirement, a new baby, vacation, and Christmas. In the

Table 1.3   *Happy and Sad Life Events*

Happy Events	*Sad Events*
Divorce	Death of spouse
Marriage	Death of close family member
Retirement	Personal injury or illness
Marital reconciliation	Jail term
Pregnancy	Fired at work
New baby	Change in health of family member
Buying new home	Sex difficulties
Change in work responsibilities	Change in financial state
Outstanding personal	Death of close friend
achievement	Child leaving home
Begin or end school	Trouble with in-laws
Vacation	Trouble with boss
Christmas	Minor violations of the law
	Loan less than $10,000

Source: Based on Thomas H. Holmes and Richard H. Rahe, "The Social Readjustment Rating Scale," *Journal of Psychosomatic Research* 2 (1967): 213–218.

right column are life events that are typically considered sad, such as the death of a spouse or friend, a jail term, being fired from a job, minor violations of the law, and trouble with in-laws. Surprisingly, the happy events from the left column were found to be comparable in stress value to the sad events in the right column. There are two important messages for emergency responders in this table.

First, stress comes from not just the negative or sad events that happen in life, but from the positive or happy events as well. Knowing that a serious personal injury or illness ranks nearly equally to the stress of getting married can change our perspective on sources of stress considerably. Remember that stress is defined as any demand on our body for change. All events that affect our life routines have the ability to produce stress, if we let them.

> *On the Lookout:* Remember, stress comes from not just the negative or sad events that happen in life, but from the positive or happy events as well.

Scheduling Stress-Producing Events

Second, take a close look at the actual Holmes and Rahe stress survey in Figure 1.6. Notice that some of the events listed are happenings that we can control; others are events we cannot schedule for a time when they would be convenient. We certainly don't schedule a car crash or the death of a spouse, friend, or family member. But we probably are able to schedule when we get married, change jobs, buy a new home, or make a significant change in our daily schedule. The ability to schedule stress-producing life events is particularly important for the emergency responder. When we total cumulative life stressors, there is a point for every person beyond which he or she will not be able to function if given any additional stress. We might refer to this point as the "breaking point," or the point at which any added stress in our lives will cause us to "snap."

> *Key Point:* Some of the events listed in Figure 1.6 are happenings that we can control; others are events we cannot schedule for a time when they would be convenient.

Call it what you may, the fact is that very high levels of stress interrupt our ability to function.

Social Readjustment Rating Scale

Rank	Life Event	Mean Value
1	Death of spouse	100
2	Divorce	73
3	Marital separation	65
4	Jail term	63
5	Death of close family member	63
6	Personal injury or illness	53
7	Marriage	50
8	Fired at work	47
9	Marital reconciliation	45
10	Retirement	45
11	Change in health of family member	44
12	Pregnancy	40
13	Sex difficulties	39
14	Gain of new family member	39
15	Business readjustment	39
16	Change in financial state	38
17	Death of close friend	37
18	Change to different line of work	36
19	Change in number of arguments with spouse	35
20	Mortgage over $10,000	31
21	Foreclosure of mortgage or loan	30
22	Change in responsibilities at work	29
23	Son or daughter leaves home	29
24	Trouble with in-laws	29
25	Outstanding personal achievement	28
26	Wife begins or stops work	26
27	Begin or end school	26
28	Change in living conditions	25
29	Revision of personal habits	24
30	Trouble with boss	23
31	Change in work hours or conditions	20
32	Change in residence	20
33	Change in schools	20
34	Change in recreation	19
35	Change in church activities	19
36	Change in social activities	18
37	Mortgage or loan less than $10,000	17
38	Change in sleeping habits	16
39	Change in number or family get-togethers	15
40	Change in eating habits	15
41	Vacation	13
42	Christmas	12
43	Minor violations of the law	11

Figure 1.6 *Holmes and Rahe's Social Readjustment Rating Scale (Source: Thomas H. Holmes and Richard H. Rahe, "The Social Readjustment Rating Scale," Journal of Psychosomatic Research 2 (1967): 216.)*

As a practical matter, emergency responders have a variable in their lives that does not exist for the rest of society—namely, their work. There is absolutely no way to predict what types of stress-producing events the emergency responder might encounter on the job. There is nothing ordinary about an "ordinary" day in the life of an emergency responder. Unlike marriage, buying a home, or changing jobs, we have no control over what emergency we might be called to during our next shift.

> ***On the Lookout:*** *Schedule controllable stress-producing life events to allow yourself room for the big unexpected variable in your life—emergency services.*

There is power in knowing what you can control. You have the power to space controllable life events over a period of time so that you allow yourself room for that big unexpected variable in your life—that is, emergency services. Use this power regularly for as long as you are involved in emergency services.

The stress survey done by Holmes and Rahe evolved into a rating scale used to predict the likelihood of developing a stress-related illness or accident. Both Holmes and Rahe and subsequent researchers were able to establish what we now recognize as a very clear relationship between stressful life events and the onset of illness or disease. If you are interested in your own score, complete the Holmes and Rahe Stress Test in Figure 1.7.

To score yourself on the Holmes and Rahe Stress Test:

- Award yourself the point value given for any event that has occurred within the past 12 months.
- If your total score is less than 150, you have only a 35 percent likelihood of developing a stress-related illness in the next 24 months.
- If you score between 150 and 300, your likelihood increases to 51 percent over the next 2 years.
- If your score is greater than 300, you are 80 percent likely to develop a stress-related illness in the next 2 years.

There are two things to keep in mind when using the Holmes and Rahe stress test: (1) We control how we perceive

Social Readjustment Rating Scale Self-Test

Rank	Life Event	Mean Value	Your Score
1	Death of spouse	100	_____
2	Divorce	73	_____
3	Marital separation	65	_____
4	Jail term	63	_____
5	Death of close family member	63	_____
6	Personal injury or illness	53	_____
7	Marriage	50	_____
8	Fired at work	47	_____
9	Marital reconciliation	45	_____
10	Retirement	45	_____
11	Change in health of family member	44	_____
12	Pregnancy	40	_____
13	Sex difficulties	39	_____
14	Gain of new family member	39	_____
15	Business readjustment	39	_____
16	Change in financial state	38	_____
17	Death of close friend	37	_____
18	Change to different line of work	36	_____
19	Change in number of arguments with spouse	35	_____
20	Mortgage over $10,000	31	_____
21	Foreclosure of mortgage or loan	30	_____
22	Change in responsibilities at work	29	_____
23	Son or daughter leaves home	29	_____
24	Trouble with in-laws	29	_____
25	Outstanding personal achievement	28	_____
26	Wife begins or stops work	26	_____
27	Begin or end school	26	_____
28	Change in living conditions	25	_____
29	Revision of personal habits	24	_____
30	Trouble with boss	23	_____
31	Change in work hours or conditions	20	_____
32	Change in residence	20	_____
33	Change in schools	20	_____
34	Change in recreation	19	_____
35	Change in church activities	19	_____
36	Change in social activities	18	_____
37	Mortgage or loan less than $10,000	17	_____
38	Change in sleeping habits	16	_____
39	Change in number or family get-togethers	15	_____
40	Change in eating habits	15	_____
41	Vacation	13	_____
42	Christmas	12	_____
43	Minor violations of the law	11	_____
		Total	_____

Figure 1.7 *Social Readjustment Rating Scale Self-Test (Source: Based on Thomas H. Holmes and Richard H. Rahe, "The Social Readjustment Rating Scale," Journal of Psychosomatic Research 2 (1967): 216.)*

> *On the Lookout:* When using the Holmes and Rahe stress test, remember that (1) we control how we perceive and react to life events and that (2) the tool was developed for and tested with the general population, not emergency responders.

and react to life events, and the way in which we do this can dramatically affect outcomes. (2) The tool was developed for and tested with the general population, not emergency responders. It is likely that the unique aspects of the emergency responder may produce different results than is the case for the general population.

DIFFERENT RESPONSES TO THE SAME EVENT

The final piece of the puzzle needed to understand stress has to do with why different people are affected in different ways by the same events. You may wonder why a certain company officer gets so annoyed when he finds dirty dishes in the kitchen sink that he erupts into an angry tirade, whereas others don't seem to think of it as a problem at all. Or you may notice that members of your squad react differently following a pediatric trauma arrest: Some become very somber and withdrawn, while others are apparently unaffected. Choice and experience are the two reasons why different people have different reactions to stress.

> *Key Point:* Choice and experience are the two reasons why different people have different reactions to stress.

Role of Choice

Choice is an important word to remember when thinking about stress. If we define stress as any demand for change placed on our mind or body, then our own interpretation of an event is critical in transforming it from a life event into a stressor. With every event that occurs in our lives, we make choices. We choose how to perceive the event and we choose how we will react. That is one explanation for why different people have different reactions to like events: They each choose what the event means to them and then decide how to react.

> *Key Point:* We choose how to perceive the event and we choose how we will react.

You probably work with someone whom everyone considers a happy person. Happy people seem to always be in a positive, upbeat mood. Perhaps you are one of those people. By choice, you certainly could be. We all make choices every day about how we are going to react to others and how we will behave. Happy people have no less stress in their lives than do the rest of us; they simply choose to be happy. It will come as no surprise to you then that miserable people choose to be miserable.

Role of Experience

Experience is also an important element in why different people react differently to similar events. One of the most dramatic studies of differing levels of stress based on experience was conducted on sky divers by Fenz and Epstein in 1967 [7]. The two researchers compared self-reported levels of stress between experienced and novice sky divers, beginning the night before a planned jump through their landing on the ground with their parachutes.

The differences between experienced and inexperienced sky divers are dramatic, as illustrated in Figure 1.8. At most times, their stress levels are quite opposite, with the major difference

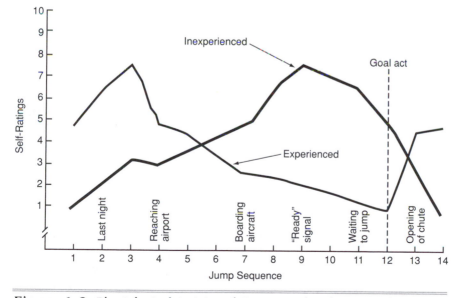

Figure 1.8 *Physiological Activity of Experienced and Novice Sky Divers (Source: Walter D. Fenz and Seymour Epstein, "Gradients of Physiological Arousal in Parachutists as a Function of an Approaching Jump," Psycho-somatic Medicine 29, No. 1 (1967): 34.)*

being that experienced sky divers seemed able to control their stress responses as they needed in order to maximize their performance during the jump as well as to increase their enjoyment during and afterwards.

Sky divers are much the same as emergency responders. Both voluntarily choose to participate in dangerous activities and enjoy doing so. People who try to find adventure in activities that others in society might consider dangerous are often referred to as "thrill seekers." Emergency responders are all inexperienced at the start. It stands to reason that novice responders, like the novice sky divers, are likely to experience greater levels of stress or anxiety as they respond to a call and during the initial efforts at the scene. Veteran fire fighters, medics, or cops, in contrast, probably experience opposite stress levels in a fashion similar to that of the experienced sky divers. As emergency responders, we should recognize that experience and length of service may significantly affect the level of stress a responder feels.

SUMMARY

Stress is a response our body makes to a demand for change. Life events, people, animals, and events make demands on us for change. Stress is our reaction to these outside influences. Although we tend to associate stress with negative life events, research has shown that positive or happy occurrences can be just as stress producing as negative or sad events.

We need a certain level of stress to be productive and enjoy life. People completely without stress are all dead. The link between too much stress and illness or disease has become increasingly clear in modern medicine. Nearly 80 percent of health problems are thought to be caused, or aggravated by, overstress [4]. Emergency responders must be able to recognize life events that can lead to stress.

The power to control stress can be as simple as scheduling life events that are within our capability to manage, such as changing jobs, getting married, buying a new home, or significantly modifying a daily routine. It is essential for the emergency responder to allow space in life for unexpected stressors that occur on the job. Spacing out controllable life events ensures that the emergency responder will be ready and able to accommodate unpredictable spikes in stress at work.

Choice and *experience* are key terms that explain why different people react in different ways to similar stressful events. When an event happens in our lives, we choose how to perceive the event and how we will react to it. Happy people choose to be happy; miserable people choose to be miserable. Studies on sky divers demonstrate that experience allows them to control their stress response in order to maximize performance and increase their enjoyment. Novices have stress levels that are often opposite to those of experienced divers. Different levels of experience likely produce different levels of stress among emergency responders as well.

REFERENCES

1. Yerkes, Robert M., and Dodson, John D., "The Relation of Strength of Stimulus to Rapidity of Habit Formation," *Journal of Comparative Neurology and Psychology* 18 (1908): 459–482.
2. Selye, Hans, *Stress Without Distress,* New York: Harper & Row, 1974.
3. Cannon, Walter B., *The Wisdom of the Body,* 2nd ed., New York: W.W. Norton, 1939.
4. Seaward, Brian Luke. *Managing Stress: Principles and Strategies for Health and Wellbeing*, 3rd ed., Sudbury, Massachusetts: Jones and Bartlett Publishers, 2002.
5. Curling, Thomas B., "On Acute Ulceration of the Duodenum in Cases of Burn," *Médecine et Chirurgie,* Translated 1842; 2nd series: 25: 260–281.
6. Holmes, Thomas H., and Rahe, Richard H., "The Social Readjustment Rating Scale," *Journal of Psychosomatic Research* 2 (1967): 213–218.
7. Fenz, Walter D., and Epstein, Seymour, "Gradients of Physiological Arousal in Parachutists as a Function of an Approaching Jump," *Psychosomatic Medicine* 29, No. 1 (1967): 33–51.

Traits of Emergency Responders

2

Certain traits or aptitudes make a good fire fighter, a good law enforcement officer, or a good medic. Just what is the "right" personality for emergency services? What are the traits these individuals should have? Emergency responders need, of course, to be cool, calm, and collected individuals—people who can perform well under pressure. The unpredictable nature of emergency calls would seem to require that responders be innovative and flexible as well. Are the traits of the typical emergency responder any different from those of the general population?

ROLE OF ADRENALINE

Many fire and police departments use psychological testing, interviews, and other tools to screen applicants. The emergency services have become very sophisticated in their ability to screen out individuals whose behaviors and characteristics are undesirable. These include people who are prone to violent or abusive behavior, drug use, alcoholism, dishonesty, or thievery. With so much known about traits and characteristics that make for a bad or unsuccessful candidate in the emergency services, surprisingly little scientific evidence exists about what characteristics make for a good emergency responder.

It does seem that emergency responders are indeed very different from the average citizen. It is not at all uncommon for family, friends, and neighbors to praise emergency responders. A certain awe and reverence for the work that emergency responders do is often based on the belief held by average citizens that they themselves would not be able to handle the stresses of emergency services work.

This belief is well founded. It may come as a surprise to emergency responders that roughly 90 percent of the population lacks the personality characteristics necessary to perform well in emergency work [1].

People who tell you that they "can't stand the sight of blood" or that they "would never be able to crawl into a burning building" are being truthful. These individuals simply don't have the requisite personality traits to do these things. Fortunately, what 90 percent of the general population lack, the other 10 percent have. Those 10 percent are often attracted to work in emergency services, and thankfully so.

One of the earliest descriptions of the emergency services personality appeared in a 1986 *Firehouse* magazine article by Jeff Mitchell, an assistant professor of Emergency Health Services at the University of Maryland. In it, Mitchell characterized the typical personalities of police officers, fire fighters, and paramedics using a standard personality inventory [2]. Although his final results were never published, his early descriptions have been widely quoted and consistently recognized as the personality framework for numerous high-stress occupations, including emergency services (see Figure 2.1).

The major difference between emergency responders and the general public can be summed up in one word: adrenaline.

> **Key Point:** *The major difference between emergency responders and the general public can be summed up in one word: adrenaline.*

Adrenaline, a hormone secreted in response to any alarm that sounds within the body—whether real or imagined—is the body's chemical stress response system. In emergencies, adrenaline stimulates the body to respond: The heart beats faster and more forcefully, blood pressure increases, and breathing quick-

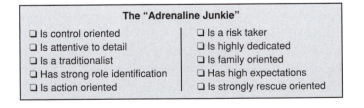

The "Adrenaline Junkie"	
❏ Is control oriented	❏ Is a risk taker
❏ Is attentive to detail	❏ Is highly dedicated
❏ Is a traditionalist	❏ Is family oriented
❏ Has strong role identification	❏ Has high expectations
❏ Is action oriented	❏ Is strongly rescue oriented

Figure 2.1 *Personality Framework for High-Stress Occupations (Source: Based on Jeffrey T. Mitchell, "Living Dangerously: Why Some Firefighters Take Risks on the Job," Firehouse 11, No.8 (1986): 50–51, 63.)*

ens. The net result of adrenaline rushing into the bloodstream is a speedy energizing of the major muscles needed to fight or run, and extra blood flow to the brain to increase alertness, reaction speed, and mental abilities.

Imagine a driver on a dark rainy night, barreling down a country road when suddenly, without warning, a deer appears in the roadway. Swerving to miss the deer, the car skids sideways, and spins around several times on the asphalt. Having missed the deer, but lost control of the car, the driver struggles to get the vehicle back on the road. After overcorrecting into another skid, the car finally comes to a safe halt at the side of the roadway. The average citizen would probably shake with terror for 20 minutes before being able to continue home. Once home, he or she might even put the car away for a while before getting up the nerve to drive again. The typical emergency responder in the same situation would probably categorize the entire experience as "fun" and wish to do it all over again. Simply put, emergency responders are happy when their adrenaline is flowing; average citizens are not.

> **Key Point:** *Emergency responders are happy when their adrenaline is flowing; average citizens are not.*

No doubt, most emergency responders will probably identify with many of the characteristics listed in Figure 2.1, the "Adrenaline Junkie." Keep in mind that although quite a few of these characteristics are individually seen in the general population, people who possess many of them together are likely to belong to the roughly 10 percent of society that has what it takes to be successful in the emergency services.

TRAITS COMMON TO EMERGENCY RESPONDERS

Control Orientation

Emergency responders tend to be control-oriented people. Their job is to create order out of chaos, making snap assessments and swift decisions, and taking quick action. They are problem solvers or people who "know exactly what to do." The emergency services culture cultivates control orientation by promoting those most

> **Key Point:** *Emergency responders tend to be control-oriented people. Their job is to create order out of chaos, making snap assessments and swift decisions, and taking quick action.*

successful at control to leadership positions. Control-oriented people are an interesting breed. Their desire to take charge often makes them difficult to supervise. They do not accept criticism well, nor are they happy when the results of their efforts are unsuccessful. Some control-oriented people have difficulty working as members of a team. Yet these same folks excel at stepping into a crisis, calming others, and proficiently resolving emergencies. Efforts at control extend internally as well. Emergency responders become expert at controlling displays of emotion. To some, asking for help with a task or displaying emotion during or after an incident is considered a sign of weakness.

Attention to Detail

Attention to detail is a common characteristic of emergency responders. "A place for everything and everything in its place" might be considered the motto of detail-oriented people. The ability to meticulously organize equipment and supplies is very beneficial in emergency services (see Figure 2.2). Having the right tools for the job at hand is critical in an emergency. Likewise, the ability to pick up on small details or clues can have enormous importance for

> **Key Point:** *The ability to meticulously organize equipment and supplies is very beneficial in emergency services.*

a critically injured patient or the safety of others at a violent crime scene or while operating at a major fire.

The admonition to avoid "tunnel vision" at emergency scenes needs to be continually emphasized. Tunnel vision means paying attention to small details at the expense of ignoring the bigger, overall picture. Since emergency responders are more commonly attentive to detail, they need to frequently remind themselves to deliberately step back at times and adjust their focus from the smaller details to the big picture of the whole scene.

> **On the Lookout:** *Avoid "tunnel vision" at emergency scenes and deliberately step back at times and adjust your focus from the smaller details to the big picture of the whole scene.*

Traditionalism

Emergency responders are often traditionalists. A sense of tradition—that is, knowing what will happen and how it will be

Figure 2.2 *Well-Organized Compartment Storage of Fire Truck (Source: Courtesy of Emergency One, Inc.)*

done—offers stability and comfort. It may be that traditional equipment such as red fire apparatus, white ambulances, and black leather gun belts offer an anchor to emergency responders who so often find themselves in bizarre and unpredictable situations. Department routines and operating procedures offer clear guidance and obvious expectations to members. Difficulties develop when new and different situations appear that call for change or adaptation. Tradition resists change, and people with traditional ways of thinking and acting often have difficulty accepting change. For a department, the inability to change with the evolving needs of the community can spell big trouble. Every department has its dinosaurs. They typify the traditionalist trait of many emergency responders.

Strong Role Identification

Emergency responders often have strong role identification. The twenty-four-hour, seven-day-a-week aspect of the job, coupled with a perceived social prestige, make it desirable to be a fire fighter, police officer, or medic. Emergency responders are proud of what they do, and rightfully so. A strong identification with their role in emergency services leads them to carry a badge or other identification even when off duty. Many wear a department jacket, patch, or job shirt and display decals or other markings on their personal vehicles to identify themselves as member of their organization (see Figure 2.3). Emergency responders like recognition. They enjoy seeing themselves in newspapers and television. Public recognition of their efforts and contributions is especially cherished. Strong role identification is uncommon in most other professions, and most individuals in the general population actually shy away from public attention.

> **Key Point:** *A strong identification with their role in emergency services leads members to carry a badge or other identification even when off duty.*

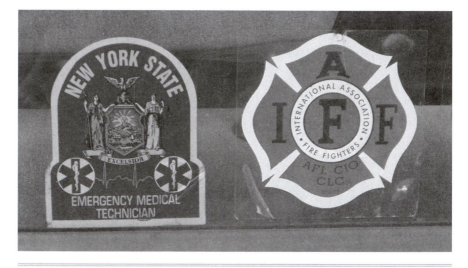

Figure 2.3 *EMT and Fire Fighter Decals on a Vehicle*

Action Orientation

Emergency responders tend to be action-oriented people. They seek out calls that sound exciting and might respond even when not dispatched or when they're off duty. They also have a strong tendency to pursue action-oriented activities outside of work. Emergency responders are more likely to ride motorcycles, skydive, race cars, engage in competitive sports, or pursue other action-packed interests rather than stamp collecting, wood working, and model airplane building for recreation (see Figure 2.4). Action-oriented and dangerous activities provide the same adrenaline high that emergency responders enjoy from "big calls" at work.

Emergency responders probably have a greater need for stimulation than the average citizen. The typical cop, fire fighter, or medic is easily bored. When things are slow or nothing is happening, they are known to initiate their own activi-

Figure 2.4 *New York City Fire Department Versus New York City Police Department Hockey Game (Source: Courtesy of the New York City Fire Department)*

ties, engaging in pranks, practical jokes, exercise, chores, or other active pursuits, whether constructive or destructive.

Risk-Taking Behavior

Risk-taking behaviors are frequently seen in emergency responders. Action-oriented people are also risk takers. The emergency services lay no exclusive claim to risk taking. Mountaineers, rock climbers, and many other athletes routinely exhibit risk-taking behaviors. The tendency to take risks is important to recognize in oneself and others because risk taking requires calculated judgment. In emergencies, it is quite possible that human interest aspects of a rescue might impair objective judgment enough that a rescuer could take unacceptable risks. For example, a rescuer called to assist a small child may be affected by thoughts of his own children while another responder who encounters an elderly individual might find herself continually reminded of her own aging parent (see Figure 2.5). This is especially true with control-

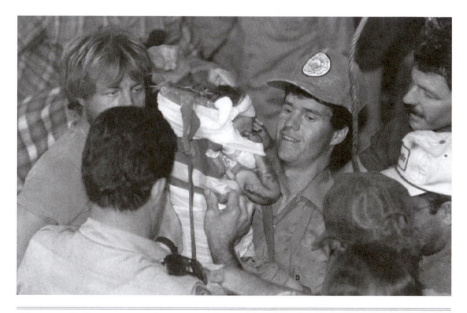

Figure 2.5 *Texas Girl Rescued from Abandoned Well (Source: AP Photo/Eric Gay)*

oriented individuals, who
work hard to keep emotions
in check; they may inadver-
tently suppress protective
feelings such as their own
fear. Checks and balances,

> ***On the Lookout:*** *Keep in mind that in
> emergencies, the human interest aspects of a
> rescue might impair your objective judgment
> and cause you to take unacceptable risks.*

like the use of safety officers, are extremely helpful to protect
the safety and welfare of people prone to risk-taking behaviors.

Dedication

Emergency responders are by and large highly dedicated folks.
They typically show intense loyalty to each other, their leaders,
and their departments. This intense dedication leads many to
consider their occupation a lifelong career. They are driven to
finish what they start and may require considerable coaxing to
leave an emergency scene even when obviously exhausted (see
Figure 2.6). Their commitment to work may lead them to spend
more and more time on the job and less time at home.

Family Orientation

In addition to their dedication to their work, emergency responders are often family-oriented people. Despite work and
on-call schedules that commonly lead to missed family gather-
ings and meals, odd sleep schedules, holidays away from home,
and the like, two strong family elements are often seen in
emergency services. First, careers in police work, fire fighting,
and emergency medical services often run in families. Many fa-
thers, sons, daughters, brothers, uncles, and cousins are fire
fighters or cops. This is tradition, and it is family. Second,
coworkers become an extended family for the emergency responder. The same odd work schedules that interfere with
home life lead emergency responders to socialize and spend
recreation time with coworkers who share their unusual work
schedules.

There is also a strong brotherhood (and sisterhood) that
unites all emergency responders. When a fire fighter, cop, or
EMS provider dies in the line of duty, the parade of brothers
and sisters that come to pay their last respects to their fallen
comrade is enormous (see Figure 2.7). If an attorney or sales-

Figure 2.6 *Exhausted Fire Fighter at Emergency Scene (Source: © 2001 FDNY Unit. All Rights Reserved.)*

person died on the job, would the same contingent of fellow workers come to pay their respects? Not likely. Emergency responders belong to a worldwide family of brothers and sisters who share their mission, their joys, their sorrows, and their pains.

Figure 2.7 *Fire Fighters at Memorial Service Honoring Six Worcester, MA, Fire Fighters (Source: AP Photo/Winslow Townson)*

High Expectations

Emergency responders tend to have very high expectations. They expect the most from themselves and from each other. There is little room for mistakes during emergency operations, but mistakes occur. Emergency responders often blame themselves for errors, and can be extremely critical of their own performances, even when things go very well. Overly high personal expectations can lead to the negative consequence of poor self-esteem. Sometimes, responders extend their high expectations to others as well. They may be quick to criticize or play Monday-morning quarterback to the actions of their coworkers.

As an industry, the emergency services repeatedly replay past performances at incidents in order to critique for errors. High expectations are good when they lead to high performance. After all, repetitive training, drilling, and practice lead to excellent performance. Perfectionism, however, is never good. No individual is per-

> *On the Lookout:* *To help maintain a positive atmosphere in the department, avoid perfectionism.*

fect, and mistakes will inevitably occur when people are involved. Avoiding perfectionism helps to maintain a positive atmosphere in the department.

Rescue Orientation

Finally, emergency responders more often than not are strongly rescue-oriented people. They would be more likely than the average citizen to stop and help a stranded motorist or come to the aid of a fallen neighbor. This makes sense, given that emergency responders often see helping others as a full-time job. Rescue-oriented people dislike calls where their special skills and abilities are not required— such as false alarms, "smells and bells," or the, "I've fallen and can't get up" calls. They become even more frustrated when opportunities to utilize their expertise fail to materialize.

NEGATIVE ASPECTS OF EMERGENCY RESPONDER TRAITS

Successful emergency responders probably possess many if not most of the traits just described. People with these qualities are usually drawn to work in the emergency services. In general, all of these personality characteristics help people succeed in emergency services work. A few of these traits, however, require the individual to exercise caution in order to avoid problems with self-esteem or relationships. Two of the emergency responder characteristics that are helpful at work may be harmful away from the emergency scene; these are control orientation and high expectations.

The Challenge of the Control Orientation Trait

Control-oriented people prefer to be in charge and are fiercely independent thinkers. They often don't take direction well, and are difficult to supervise. In fact, control-oriented people comprise a significant portion of folks labeled as "difficult people." [3] Many emergency responders freely admit that they consider themselves to be difficult people. Their control orientation explains why this is so.

Although being control oriented is useful for operating effectively at an emergency scene, control-oriented people often experience difficulties on the job and in their relationships with others. Not surprisingly, nearly all troubles at work revolve around issues of control. When there are problems at work as well as difficulties with relationships at home, the emergency responder may become isolated. Isolation is the most common cause of stress at home for the emergency responder. Sadly, being deprived of meaningful contact with those we love adds even greater stress to our lives. Recognizing that you are a control-oriented individual is the first step in reducing the problems that often result.

> **Key Point:** *Isolation is the most common cause of stress at home for the emergency responder.*

> **On the Lookout:** *Recognizing that you are a control-oriented individual is the first step in reducing the problems that often result.*

The Challenge of the High Expectations Trait

High expectations help to keep emergency responders prepared for the many different situations they encounter. At extremes, people with high expectations become perfectionists. Continually expecting perfection from ourselves leads to poor self-esteem while demanding perfection from others leads us to become overly quick to criticize. Poor self-esteem and continued criticism of others involve a whole lot of negative thinking and will often become a self-fulfilling prophecy.

Emergency responders can prevent problems from high expectations by teaching themselves to listen for negative thinking and choose instead to focus on the positive. This is particularly important in the department or station when other emergency responders are around. Negativity among a group of emergency responders (who all have some degree of high expectations) is contagious and will spread like a virus.

> **On the Lookout:** *Avoid problems from high expectations by teaching yourself to listen for negative thinking and choosing to focus, instead, on the positive.*

SUMMARY

Emergency responders are unique people, the likes of which are found in only about 10 percent of the population. A take-charge attitude, a desire to do things in an organized and detailed manner, a certain degree of fearlessness, a high degree of dedication to the job, respect and concern for fellow members of the emergency services family, and the highest standards and expectations—all these qualities combine to produce exactly what the public needs in an emergency response. The combination of personality traits that draw people to work as emergency responders also help them succeed and can be summed up as enjoyment of the "adrenaline rush."

Traits such as control orientation and high expectations that help emergency responders can also significantly hinder them. An awareness of the behaviors associated with these traits can help emergency responders avoid personal and professional consequences such as isolation and persistent negativity. A full awareness of the differences between people drawn to work in the emergency services and the remainder of the population can be useful for both emergency responders and their significant others in understanding the forces that motivate them and the potential traits that may lead to difficulties in relationships with both coworkers and family.

REFERENCES

1. Hopper, Linda, "Stress Recovery: An Interview with Jeffrey Mitchell, Ph.D.," *Public Management,* November, 1988: 5–8.
2. Mitchell, Jeffrey T., "Living Dangerously: Why Some Firefighters Take Risks on the Job," *Firehouse* 11, No. 8 (1986): 50–51, 63.
3. Bacal, Robert, *Complete Idiot's Guide to Dealing with Difficult Employees,* New York: Alpha Books, 2000.

Stress on the Job

3

It takes a distinct set of personality traits to be successful in the emergency services. People with these traits are probably attracted to the stressful nature of the emergency responder's job. Some of the traits that help make a good fire fighter, cop, or medic also hinder work relationships and may actually increase job stress. This chapter explores stress on the job with an eye on how the emergency responder can make the workplace more enjoyable. No one looks forward to going to work in a hostile, unfriendly, or negative department. Worse yet, people who don't enjoy their jobs eventually find their dissatisfaction spilling over into their personal lives. Emergency responders can take specific measures to keep themselves happy at work and at home.

OCCUPATIONAL STRESS VERSUS JOB STRESS

Work as an emergency responder involves two distinctly different kinds of stress: occupational stress and job stress. Occupational stress pertains to the specific influences of emergency response work and is covered in detail in Chapter 4. Job stress is stress related to the work environment, the boss, and relationships with coworkers. Emergency responders find enjoyment in occupational stressors, but few welcome job stress.

As discussed in Chapter 2, most job stress in emergency services revolves around issues of control. A second factor contributing to job stress for emergency responders is high expectations. Control orientation and high expectations, as previously noted, are two personality traits commonly observed in emergency responders.

Unlike most books on stress, we have not selected a specific psychological model of stress or a particular program to help you reduce stress and begin living well. Sigmund Freud, Carl Jung, Elisabeth Kübler-Ross, Viktor Frankl, Leo Buscaglia, Abraham Maslow, Gail Sheehy, and others have written volumes and inspired others to produce countless additional books with theories on stress and recommendations for living well in a stressful world. Stress management theories and models are like pants: No one pair fits everyone.

Similarly, no stress management theory fits everyone, and no stress management model works against all types of stressors. If you want to reduce your stress by improving your diet; practicing better time management; improving your physical fitness; or learning meditation, yoga, hypnosis, or biofeedback techniques, there are thousands of books out there to help you. Rather than choosing a specific stress management theory or model, we choose instead to focus here on how the personality of emergency responders can increase stress on the job. With a better understanding of on-the-job stress comes an improved ability for emergency responders to take charge of their own happiness. In Chapter 8 we'll apply the concepts introduced in this book to helping you choose a personal stress management model.

SYMPTOMS OF OVERSTRESS

In Chapter 1 we saw that stress affects different people in different ways. When stress becomes a negative influence, a person is more susceptible to injuries and illness, and usually does not enjoy an overall sense of well-being. This uncomfortable state of distress, or being harmfully influenced by stress, is called *overstress*. Before examining the specific influences of control and high expectations on the emergency services workplace, reviewing some of the symptoms of overstress might be beneficial. Not everyone has the ability to know when his or her personal stress meter is rising. When individuals are unable to recognize stress-induced behavioral changes, they may need the help of coworkers or friends to make them aware of their symptoms. Common flags that point to very high levels of stress include the following:

- Frequent illnesses or absences from work
- Sleep disturbances

- Increased use of alcohol or drugs
- Overeating
- Frequent fights with others
- Relationship troubles
- Depression
- Avoidance or quitting behaviors
- Increased mistakes on the job

When you see these behaviors in yourself or others, you should ask, "Could this be a symptom of overstress?"

In mainstream society, job stress is at an all-time high, far exceeding all other stressors in people's lives. The costs of job stress are monumental in both dollars and lost productivity. One study estimated the costs of stress to employers in the United States at an average of $200 billion per year. Some 60 to 80 percent of industrial accidents have a stress-related cause, as do 80 percent of all visits to primary care physicians [1]. According to a list compiled by the National Safety Council, the cause of all this stress is lack of control. Emergency responders enjoy no immunity from the stressors that affect the rest of society. In fact, responders may actually experience greater levels of job stress as a direct result of their unique personality traits.

> ***On the Lookout:*** *When you see behaviors such as frequent illnesses or absences from work, sleep disturbances, increased use of alcohol or drugs, overeating, frequent fights with others, relationship troubles, depression, avoidance or quitting behaviors, or increased mistakes on the job in yourself or others, you should ask, "Could this be a symptom of overstress?"*

CONTROL ORIENTATION AS A SOURCE OF CONFLICT

As discussed in Chapter 2, control orientation is a personality trait common to emergency responders and is a necessary trait for them because they need to take charge quickly at emergency scenes. The public expects emergency responders to roll in with a take-charge attitude and quickly gain control of whatever crisis they encounter. Control-oriented people work well independently. Their ability to control their own emotions helps them remain calm and functional even under tremendous duress. When things go as planned, control-oriented people are happy and pleased with the effects of their efforts. When things go poorly or not as planned, control-oriented people often perceive themselves as having failed to control the situation and may be extremely displeased or unhappy with both themselves and others involved.

In a typical workplace, there might be one or two control-oriented people. Usually these individuals would be the managers or supervisors. In contrast, a firehouse, police station, or ambulance squad is full of control-oriented people, a situation that is a recipe for conflict. Control-oriented people prefer to be in charge, not be told what to do. Yet successful responses to emergencies require that one person be in charge. Anyone with a few years of experience in public safety has witnessed scenes gone horribly wrong when there were too many generals and not enough soldiers. Realizing that every emergency responder is control oriented to some degree is important both at emergency scenes and in the daily workplace. The control orientation trait of emergency responders explains why a great deal of conflict occurs on the job.

CONFLICT MANAGEMENT

In a work environment like emergency services, the presence of many control-oriented individuals has the potential to increase conflict between people. For that reason, every emergency responder from the chief right down to the newest member in the department must have skills in conflict management. Knowing how to resolve conflict won't prevent it from occurring, but it will go a long way toward preventing conflicts from becoming significant job stressors for the individuals involved.

Conflict in emergency services wears many faces. Chiefs may feel unable to control their budget and conflict with government officials. Paramedics may not like the placement of equipment in their department ambulances and conflict with superiors whom they are unable to convince of the need for change. Fire fighters may conflict with department scheduling officers whom they can't convince to change a holiday schedule. Police officers may conflict with fellow officers over whether or not to make an arrest in a family dispute (see Figure 3.1).

Figure 3.1 *Police Officer Mediating Argument Between a Couple (Source: © Richard Lord/The Image Works)*

The styles used to manage these conflicts can be either helpful or lead to additional problems. Unresolved conflict or conflict resolved in a manner that does not allow both parties to feel in control of the ending will result in anger and frustration that can eventually lead those involved to burnout.

Negative Ways to Manage Conflict

Three negative ways to manage conflict are by withdrawal, surrender, or hostile aggression [2].

Withdrawal

In withdrawal as a negative way of managing conflict, the person walks away from or avoids the situation or otherwise remains silent instead of attempting to resolve the conflict. Withdrawal leads to anger and resentment and is often found in situations where one party feels outnumbered or outgunned by authority figures. Withdrawal used to "cool off" can also be a positive style of conflict management, however, if it is followed by attempts to resolve conflict.

Surrender

Surrender is another type of withdrawal in which a person habitually gives in to the other person in order to avoid conflict. The "Yes, dear" response is a classic example of surrender seen in relationships. Surrender can lead to resentment and feelings of victimization and can actually prevent a resolution of the conflict.

Hostile Aggression

The hostile aggression response turns conflict into confrontation. Often verbal and sometimes physical aggression is used to intimidate, control, or manipulate others into passive agreement. Aggression fails to resolve conflict and often breeds deep resentment. When permitted, aggressive behavior can become a cyclic pattern used continually to manage conflicts.

> *Key Point:* Three negative ways to manage conflict are by withdrawal, surrender, or hostile aggression.

Each of these negative conflict management styles leaves one or both parties having lost the battle. When either party walks away without having resolved the conflict, increased stress is likely to result.

> **Case:** A notice from the Chief appeared without warning on the bulletin board at Station 7. In what seemed to be a never-ending series of surprises, Chief Green had abruptly decided to change the work shifts of several fire fighters. Marie, José, Karen, and Bob were four fire fighters at Station 7 who abruptly had their lives turned upside down by this Friday afternoon memo.
>
> Bob was the most vocal with his displeasure. In fact, vocal would be understating Bob's reaction. A fire fighter for 5 years, Bob had a reputation for throwing temper tantrums when things weren't going his way. Within seconds, Bob was on the phone with the Chief's secretary, cursing up a storm. When he couldn't get the Chief himself on the line, he slammed the phone down, tore the receiver off the wall and hurled it out into the apparatus bay. Turning, Bob stormed out the back door, slamming the door so hard that it shook the station. [*Bob uses hostile aggression.*]
>
> "Wow, that was dramatic," Karen said to Marie, who, along with several other fire fighters, had witnessed Bob's

tirade. "I can't really see how that kind of behavior is going to get him anywhere," Karen continued. "In fact, the sooner he realizes that the administration is out to get us, the better off he'll be. I recognized a long time ago that there's nothing we can do to make our situation any better around here. We're just basically victims of every new change that comes down the pike. Bob needs to just give in . . ." [*Karen uses surrender, a form of withdrawal.*]

José interrupted, "I'm just about as steamed as Bob, and I'm half tempted to drive right over to the Chief's house and give him a piece of my mind about this. I'm afraid though, that I might act irresponsibly so I'm going to forget about the whole thing for a couple days. After I cool off, though, I'm going to sit down with the Chief and try to work this problem out." [*José uses withdrawal to "cool off," planning to resolve the problem later once he's had an opportunity to calm down.*]

"I refuse to even think about it at all," Marie blurted out. "There's no way I could ever win a battle with the Chief, so I don't even see the point of trying. I'm not going to think about it again." [*Marie uses withdrawal.*]

Positive Ways to Manage Conflict

The two positive ways people use to manage conflict are through open dialogue and persuasion [2].

> **Key Point:** *The two positive ways people use to manage conflict are through open dialogue and persuasion.*

Open Dialogue

In open dialogue, attempts to reach a settlement everyone can agree on are made through discussion and negotiation. Dialogue often requires compromise. It involves discussion of the pros and cons of resolving the problem. Gaining a better understanding of how each person views a problem is key to resolving conflict through open dialogue.

Persuasion

Persuasion is used to diplomatically suggest new ideas in order to resolve conflict and encourage different attitudes or new ways of viewing a problem.

Both of these positive conflict management styles leave all sides with a "win-win" ending. No party walks away feeling that he or she lost control over the outcome.

Knowing When to Let Go

Sometimes circumstances arise that the emergency responder cannot control and no amount of conflict management skill or training will be of benefit. All of us have probably had to work with or at least have met people who are monsters. We may also have found ourselves from time to time trapped on a shift, with a boss, or in a predicament that we simply didn't like at all. Such circumstances require a third style of positive conflict management—that is, knowing when to let go. In a conflict situation, the responder must decide whether it is something he or she *can* control (see Figure 3.2). If not, then he or she must let it go. Trying to influence people or things that are beyond our capability to influence can result in anger, resentment, and frustration. All of these increase emergency responders' stress.

Letting go of a conflict is acceptance, not withdrawal or surrender. The difference between these is what makes letting go a positive conflict management style. Both withdrawal and surrender avoid dealing with conflict by either walking away or simply giving in. In these situations, the conflict remains. Letting go requires thinking about what actually falls within your span of control. It will be obvious that some circumstances are beyond your influence or not worth the effort required to resolve them. These conflicts quite simply, should be let go right then and there, accepting that they will not change.

Most of us practice letting go every day. We may be annoyed about taxes, high fuel prices, or the way our spouse squeezes the toothpaste tube. Rather than confronting these conflicts, or losing sleep feeling resentful and victimized from avoiding them, we choose to accept them, that is, we choose letting go. We recognize that there are other problems toward which our energies can be much more constructively applied. Letting go is taking a step back and asking yourself, "What influence do I really have here?" and "Is it worth the effort required to resolve this problem?" If the answers are "none" and "no," learn to let go.

On the Lookout: *Take a step back and ask yourself, "What influence do I really have here?" and "Is it worth the effort required to resolve this problem?" If the answers are "none" and "no," it's time to let go.*

Figure 3.2 *Confrontation Scene (Source: Reproduced with permission from Emergency Telecommunicator: Course Manual, The National Academies of Emergency Dispatch (NAED), p. 174, Figure 8.9. Courtesy of Jones and Bartlett Publishers)*

HIGH EXPECTATIONS AS A CAUSE OF STRESS

As discussed in Chapter 2, the second personality trait common to emergency responders that increases job stress is high expectations. Emergency responders expect the most from themselves and from everyone around them. High expectations are normally helpful to the emergency responder, but can lead some people to become perfectionists. No person or thing can be perfect and those who expect perfection are certain to be disappointed!

Continually demanding perfection (or overly high performance) from ourselves can eventually damage our self-esteem. People who routinely demand perfection of themselves are often quick to criticize less-than-perfect performance in others. Poor self-esteem and constant criticism of others are both forms of negative thinking. A workplace loaded with negative thinking becomes a toxic environment and fosters negative behavior. Because emergency responders traditionally hold high expectations, it is particularly important for each responder to maintain a heightened state of alert for negative thinking, or for what psychologists call "toxic thoughts."

> **On the Lookout:** *Maintain a heightened state of alert for negative thinking, or for what psychologists call "toxic thoughts."*

Negative Self-Talk

The first place toxic thoughts are found is within the individual. Negative self-talk includes the following:

- Pessimism: calling the glass half empty, never half full
- Magnifying: blowing problems out of proportion
- Blaming: always assigning fault to someone else
- Perfectionism: expecting more than humanly possible
- Victimization: seeking pity or sympathy from others
- "Should"-ing: scolding yourself for things you should have done
- Self-comparison: continually assessing yourself against others

> **Key Point:** *Negative self-talk produces negative results and leads to poor self-esteem.*

Listen for these thoughts to appear, and recognize them as toxic thoughts with all the potential to become self-fulfilling prophecies. Negative self-talk produces negative results and leads to poor self-esteem. Fight negative self-talk that creeps into your mind. Choose instead to focus on your assets and abilities. This will shift your self-talk from negative to positive. Remember that you are not alone in having high expectations of yourself.

> **On the Lookout:** *Fight negative self-talk that creeps into your mind. Choose instead to focus on your assets and abilities.*

Case: Susan had been a police paramedic for the past five years. She enjoyed working in the business district of the same small city where she had been born some 30 years earlier. Lately, it seemed to Susan that she just couldn't do anything right. She frequently missed IVs and endotracheal tubes that her partner seemed to have no trouble getting in on his first try. After almost every EMS call, three or four ideas would pop into her mind of actions she should have taken, but just couldn't seem to remember at the right time. Some days, it seemed that she just couldn't do anything right. When Susan compared herself to other police medics, which she had the habit of doing regularly, she seemed to come out at the absolute bottom of the pack. [*Susan practices negative self-talk.*]

Negativity in the Department

Toxic thoughts tend also to spill out of individuals, especially in the emergency services environment where many if not most people hold high expectations. In addition to listening for negative thinking within themselves, emergency responders must listen for negativity in their department. Perfectionism and quick criticism of others are the major

> **On the Lookout:** *In addition to avoiding negative self-talk, you must listen for and counter negativity in your department. Perfectionism and quick criticism of others are the major signs of toxic thinking in a department.*

signs of toxic thinking in a department. Members either contribute to negativity or promote a positive climate every time they open their mouth.

Because there are so many people with high expectations in emergency services, negativity released into a department spreads like a highly contagious virus. In today's era that emphasizes customer service, it is important for emergency service providers to remember that their fellow members are equally important, if not more important so, than the department's customers. We each deserve the same respect and professional treatment that the public enjoys from our department. Where pessimism, toxic thinking, and negativity are not welcomed, a positive attitude exists. A positive workplace environment requires effort from each and every member.

Case: "I've about had it with this new engine," Steve proclaimed as he climbed out of the driver's seat, "I can't believe the city bought this piece of junk." "You should know how government operates by now," Rich chimed in, "Low bid; no concern for us little guys on the front line." "I can't remember a week since we got this engine that it hasn't sat in the shop for a day," Lisa added. "Too bad the Chief won't stand up to them. That might get us some decent equipment around this place." "Dream on Lisa," Steve exclaimed, "All the officers around here are looking out for themselves above everything else!" [*Example of negativity in the department.*]

STRENGTHS OF EMERGENCY RESPONDERS

In no way does this chapter intend to imply that emergency responders are affected by job-related stress any more than people in the rest of the working world or that they are less able to cope with stress. Their unique personalities, however, may make emergency responders more prone to conflict because they tend to be control oriented. Their high expectations increase the chances for negativity to spill over from individuals and spread within their ranks. Despite working in higher stress jobs, studies [3] conducted in the 1970s suggest that people with personalities like emergency responders (those inclined to sensational activities like rock climbing, windsurfing, hang gliding, and sky diving) may well be better able to cope with life's stressors [1]. It seems that the personality strengths necessary to perform in sensational situations (such as confidence, courage, self-reliance, creative thinking, and optimism) are the same strengths needed to effectively manage everyday stress. If these studies are correct, emergency responders may actually do better at handling stress than the rest of society, even though they encounter much more of it.

SUMMARY

Traits that attract people to work in emergency services also help make them effective on the job. Recent research suggests that some of these personality traits are the same strengths needed to effectively handle life and job stress. Control orientation and high expectations are two traits commonly identified in emergency responders. Both of these help responders—but can also hinder them by increasing job stress. The ability to recognize common flags of very high stress levels is important for emergency responders. Some responders don't recognize high levels of stress in themselves and need feedback from their peers to identify symptoms. Recognizing stress and having insight into particular vulnerabilities of emergency responders increases the likelihood of enjoying a happy work environment.

Withdrawal, surrender, and hostile aggression are negative conflict management styles. These lead to anger and frustration when conflicts remain or are thought to be resolved unfairly. Positive styles of conflict management include open dialogue, persuasion, and knowing when to let go. These help the emergency responder reduce stress resulting from many control-oriented people working and living together. Guarding against negative self-talk helps individuals maintain a positive attitude and high self-esteem. Toxic thoughts that spill from individuals into a department are highly contagious and preventing their spread is a duty of every department member.

REFERENCES

1. Seaward, Brian Luke, *Managing Stress: Principles and Strategies for Health and Wellbeing,* 3rd ed., Sudbury, MA: Jones and Bartlett Publishers, 2002.
2. Schafer, Walter E., *Stress Management for Wellness,* 4th ed., Stamford, CT: Thomson Learning, 2000.
3. Zuckerman, Marvin, "Dimensions of Sensation Seeking," *Journal of Consulting and Clinical Psychology* 36, No. 1 (1971): 45–52.

Critical Incident Stress 4

Public safety and medicine are unique lines of work with shared similarities. Both are problem-focused service delivery systems whose "customers" are typically not having a good day. Individuals who operate in these two fields of work have multiple opportunities to see and experience things that few people would ever encounter in their lifetime.

Emergency responders have a front-row seat for many of the darkest moments in life. Fire fighters are there with families who are devastated by the loss of their homes and the tragic deaths of their loved ones. EMS personnel are present during the last moments of life, as patients slip beyond the reach of medical talents into the hands of death. Law enforcement individuals regularly intervene in public struggles between criminals and their victims and in the private battles that occur daily in many homes.

Emergency responders also attend some of the happiest moments in the lives of others. The ability to deliver new life into the world, to rescue someone from almost certain death or disabling injury, or to succeed at stopping an innocent person from being victimized by another are truly rewarding and invigorating experiences. Unfortunately, happy endings and wonderful outcomes are far fewer than dark unhappy ones. Often, comfort and reassurance are all the emergency responder has to offer.

The steady stream of invitations to participate in the unusual events of other people's lives is part of an emergency responder's job. The personal value of these experiences is inestimable. Experience gives emergency responders new knowledge and greater strength to better cope with future events, whether they occur on the job or in their personal life. These experiences, however, come at a cost. The cost is often an emotional one and in some cases can spell disaster for the emergency responder.

ACUTE STRESS DISORDER (ASD) AND POST-TRAUMATIC STRESS DISORDER (PTSD)

Since the time soldiers first began returning from battle, there has been recognition that perfectly normal people exposed to an entirely unusual event can experience very strong emotional reactions. In 1980, the American Psychiatric Association officially recognized the behavioral and cognitive symptoms that can result from highly unusual experiences by including them under the term post-traumatic stress disorder (PTSD) in their *Diagnostic and Statistical Manual of Mental Disorders* [1]. The criteria to diagnose PTSD underwent several revisions and in 1994 a similar disorder—acute stress disorder (ASD)—joined PTSD in the diagnostic manuals. The major difference between these two disorders is the length of time that the symptoms continue to manifest themselves. The symptoms of ASD occur after a critical incident or major event and last at least two days but less than four weeks. PTSD symptoms are virtually the same as ASD ones, but their duration exceeds four weeks.

> **Key Point:** *The symptoms of ASD occur after a critical incident or major event and last at least two days but less than four weeks. PTSD symptoms are virtually the same as ASD ones, but their duration exceeds four weeks.*

In the course of their work, emergency responders are often exposed to extremely unusual situations that have the potential to be considerably upsetting. As with any normal person exposed to an entirely unusual event, emergency responders can and do develop ASD following a big call. Those who continue to experience symptoms beyond four weeks are considered to have progressed to PTSD.

ASD is a fact of life for emergency responders. New recruits should be taught to expect that they will develop ASD and perhaps PTSD from their work in emergency services. Not to warn them about this likelihood is an extreme disservice to them. Figure 4.1 spells out the symptoms of ASD and PTSD. Unless they are warned to expect these symptoms, emergency responders who begin to experience ASD might well conclude they were going crazy. Unfortunately, emergency responders, un-

---

Acute Stress Disorder (ASD) and Post-Traumatic Stress Disorder (PTSD) Symptoms

Individual exposed to a horrible event has both
❑ Experienced/witnessed death or serious injury
❑ Resulting significant emotional response/impression

Reexperiencing/Reliving (with at least one of the following)
❑ Recurrent images or thoughts
❑ Distressing dreams
❑ Flashbacks
❑ Increased distress with reminders of the event
❑ Sense of reliving the event or déjà vu

Emotional Anesthesia/Avoidance (with at least three of the following)
❑ Active avoidance of thoughts, feelings, or conversations about the event
❑ Active avoidance of activities, places, or people that remind of the event
❑ Inability to remember some important aspect of the event
❑ Significantly decreased interest in usual activities
❑ Feelings of detachment or estrangement from others
❑ Absence of emotional responsiveness or loss of enjoyment
❑ Sense of shortened future
❑ Numbness or feelings of being in a daze or dreamlike state

Persistent Anxiety (with at least two of the following)
❑ Difficulty falling or staying asleep
❑ Irritability or outbursts of anger
❑ Trouble concentrating
❑ State of hyperalertness or continual restlessness (can't stay still)
❑ Exaggerated startle response

Combination of symptoms produces significant impairment of ability to work, function socially, or otherwise cope with important life activities

ASD: Symptoms last between 2 days and 4 weeks
PTSD: Symptoms last more than 4 weeks

Figure 4.1 *Symptoms of Acute Stress Disorder (ASD) and Post-Traumatic Stress Disorder (PTSD) (Source: Based on American Psychiatric Association, Diagnostic and Statistical Manual of Mental Disorders, 4th ed., Text Revision, Washington, DC: 2000.)*

aware that they are suffering from ASD or PTSD, sometimes leave their organization, believing that the symptoms are signs of weakness that mean they are not cut out for the work.

EFFECTS OF CRITICAL INCIDENTS

Not every person is affected in the same way by stress, and the same holds true for critical incidents. Some incidents may be

Key Point: *The same reasons why different people are affected in different ways by the same stressful events also apply to critical incidents: choice and experience.*

highly upsetting to one rescuer, and have no emotional impact on another. There are several factors that might explain why this occurs, but suffice it to say that the same reasons why different people are affected in different ways by the same stressful events also apply to critical incidents—that is, choice and experience.

Choice as a Factor Influencing Reaction

Each of us has made choices over time dictating how we view the world. In making these choices, we develop certain sensitivities, or "buttons" that, when pushed or activated, raise our level of emotion. Those concerns are different for each of us, and can be seen in the range of reactions among members of a department. Certain kinds of calls bother some people and not others. Other calls may bother everyone or no one.

Experience as a Factor Influencing Reaction

Experience is the second factor influencing an emergency responder's reaction to a critical incident, and this happens in two ways:

1. New members to the emergency services commonly have their exposure to extreme stressors purposefully limited by their more experienced mentors.

2. Experienced responders tend to model coping behaviors that help new members become acclimated to stressful events and gory scenes.

Key Point: *No amount of experience prevents a responder from developing ASD or PTSD symptoms on occasion.*

This is not to say that emergency responders develop the ability to be unaffected by a grisly, grotesque, or extremely upsetting call. No amount of experience prevents a responder from developing ASD or PTSD symptoms on occasion. What experience does, however, is help emergency responders develop a sense of what to expect. They learn the following:

• What their reactions will be during and after incidents that bother them personally

- How best to handle their feelings
- How long it will typically take for them to feel like their usual selves again (two or three weeks in most cases)

CATEGORIES OF ASD/PTSD SYMPTOMS

The symptoms of ASD and PTSD fit into three distinct categories: re-experiencing or reliving the event, emotional anesthesia, and persistent anxiety.

> **Key Point:** *The symptoms of ASD and PTSD fit into three distinct categories: re-experiencing or reliving the event, emotional anesthesia, and persistent anxiety.*

Re-experiencing or Reliving the Event

Initially, emergency responders may have great difficulty getting a particularly distressing call out of their mind. They may see the faces of victims (in the form of flashbacks), smell odors from the scene, hear voices, or feel as though the incident were happening over and over again. Waking thoughts may invade dreams, and exposure to reminders of the event can increase other symptoms.

Emotional Anesthesia

Emotional anesthesia or numbness overcomes responders. They find themselves without interest in usual activities. Favorite foods, places, and people seem meaningless. They may actively work to avoid thoughts about the event as well as contact with people or places that trigger upsetting memories of the event. A void develops between the responder and others. There is a strange sense of distance, almost as though they were living in a dream. Often, the responder finds himself or herself emotionally unresponsive and unaffected by experiences that would normally bring them joy or sadness (including things like food and sex). In discussing the event with others, responders may learn that they are unable to recall some important details or happenings of the event.

Persistent Anxiety

Anxiety and nervousness invade the responder. He or she may have trouble falling asleep or staying asleep. Concentration becomes extremely difficult. Anger may come easily. Patience shortens, and it becomes more difficult to get along with others. Restlessness or an inability to stay still appears and is often accompanied by an extremely heightened state of alertness. Noises and touches may cause the responder to appear "jumpy," as the result of a greatly exaggerated startle response.

RESPONSES TO ASD/PTSD SYMPTOMS

Collectively, the symptoms of ASD and PTSD are very troubling to a person who experiences them. They significantly impair a responder's ability to function at work, and the ability to eat, sleep, and enjoy life with others. The good news is that they are temporary and usually last for less than two to three weeks. How the emergency services have handled these symptoms has changed considerably over the years.

Pre-1980 Responses

Until the late 1970s and early 1980s, emergency responders were expected to maintain a macho attitude following particularly distressing calls. Demonstration of the ability to control a situation and oneself meant ignoring emotional responses to difficult experiences. This commonplace belief, referred to as the "John Wayne syndrome," meant that the emergency responder could not possibly be hurt unless he or she had some physical wound to show for it, like a broken bone sticking through the skin.

Post-1980 Responses

By 1980, after finding a considerably higher incidence of PTSD symptoms in emergency responders than in the rest of the population, psychologists became concerned about the effects of PSTD. These findings remain true today. A 1998 study [2] of German fire fighters found an 18.2 percent incidence of PTSD compared to the National Institute of Mental Health's finding [3] of 3.6 percent incidence of PTSD in the general population of the United States. Indeed, study after study

Key Point: Emergency responders are far more prone to experience ASD and PTSD than is the rest of the population.

has demonstrated that emergency responders are far more prone to experience ASD and PTSD than is the rest of the population, probably because of the particularly difficult nature of their on-the-job experiences.

CRITICAL INCIDENT STRESS DEBRIEFING (CISD)

Accomplishment of CISD

Critical incident stress debriefing (CISD) was born in 1983, in an attempt to do something about the high incidence of PTSD in emergency responders [4]. Whether it has been effective in shortening the length or PTSD symptoms or not remains a topic of debate and continued research [5]. What CISD undoubtedly did accomplish was to focus attention on the fact that emergency responders are real people who have real emotions, and these emotions are associated with the many experiences they

> **Key Point:** *People who try to ignore the impact of extremely stressful events suffer from increased personal stress as a result.*

have in their work. People who try to ignore the impact of extremely stressful events suffer from increased personal stress as a result.

Perhaps the high rates of alcoholism, drug use, divorce, and suicide common to emergency responders are related to a culture that until fairly recently encouraged them not to recognize the impact that stressful events might have on their personal and professional lives. Today, there are a number of systems of group debriefing, all of which share similar characteristics. In addition to CISD, the *Raphael model* and *process debriefing* are two other respected techniques of group debriefing. Both are similar to CISD, with some slight differences in their organization and the core content. CISD is the model used most often in emergency services, and is the one we will discuss here.

Critical Incidents

The premise of CISD is that certain incidents—labeled "critical incidents" by Mitchell—are particularly upsetting to emergency responders. Included among critical incidents are occurrences such as the following:

- Line-of-duty deaths
- Serious line-of-duty injuries
- Emergency responder suicides
- Disasters (see Figure 4.2)
- Deaths of infants and children
- Tragedies involving friends or family members of rescuers
- Events attracting unusual media attention (see Figure 4.3)
- Serious injury or death of civilians during emergency operations (e.g., shootings, emergency vehicle collisions, etc.) [4]

Having a list of critical incidents is useful for reference but can cause responders alarm if they find themselves unaffected after responding to a call appearing on such a list. As stated earlier, however, every responder is different and different people will be affected in different ways by the same event. Cookbook approaches are, therefore, not useful in stress management because no two people are alike.

Figure 4.2 *Aftermath of a Tornado*

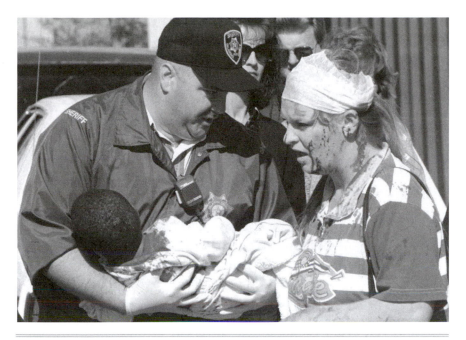

Figure 4.3 *Sheriff Department Personnel Assisting Child and Woman at Oklahoma City Bombing, 1995 (Source: AP Photo/The Daily Oklahoman, Steve Gooch)*

Department leadership must recognize that, similarly, no two incidents are alike and their members may or may not be in need of debriefing through formal CISD. Recent literature suggests that CISD (and other single-session debriefings) might not be any more helpful than no intervention and could be harmful by delaying normal recovery in some cases. Further research is needed to determine the most effective approach following a critical incident [5, 6].

> **Key Point:** *Department leadership must recognize that no two incidents are alike and their members may or may not be in need of debriefing through formal CISD.*

Departments that choose to offer CISD typically look for highly charged incidents that seem to test the coping capabilities of their responders. These incidents might include the "critical incidents" suggested earlier by Mitchell, as well as any call where responders seem very excited, upset, or agitated. Department leaders familiar with their members might notice signs of distress. These include significant mood changes, excessive humor or derogatory comments, withdrawal, or increased irritability. It's important to keep in mind that a single incident may be a critical incident for some responders on the call, but not for others. Additionally, offering CISD may be helpful and shows that leaders are concerned. However, presuming that responders need debriefing or requiring them to attend ignores individual differences and communicates lack of confidence in each responder's own coping abilities.

CISD Team

In formal CISD, the CISD team gathers a group of rescuers who have been significantly affected by an event they experienced and, following an organized seven-step process [7], the team attempts to reduce the impact of the event and expedite the return of the people involved to their normal routines. Although the CISD team is composed of mental health professionals and trained peer support personnel, the debriefing is not a therapy session but rather an opportunity for people to discuss the effects of a shared stressful experience with each other, receive some education on normal and abnormal stress responses, and devise a plan for gaining some closure on their experience. Mental health professionals are included for their expertise in facilitating groups as well as ability to recognize when an individual rescuer may need additional assistance in coping with an incident. In formal counseling or therapy, participants would be taught specific psychological and behavioral coping strategies and, although they might also discuss their stressful experience, they would do so in a gradual and systematic fashion.

> **Key Point:** *The CISD team attempts to reduce the impact of the event on significantly affected responders and hopes to expedite their return to their normal routines.*

Seven Steps of CISD

The seven-step CISD is scheduled at least twenty-four hours after an incident and may take place as long as a week later. CISD teams usually cover a regional area, serving many emergency services. Their membership is normally about one-third mental health providers and two-thirds peer support staff. To debrief 10 to12 responders, between three and five team members are needed. When large numbers of responders require debriefing simultaneously, they can be divided into smaller groups of 10 to 15 people each. A coordinator usually arranges the debriefing and pulls together information about the incident for team members. The debriefing is announced to responders involved so those who wish to attend may do so. Often times, debriefings are held in a location away from the department to avoid distractions. CISD takes between two and three hours and follows the format described here and shown in Figure 4.4.

1. *Introduction.* Team members introduce themselves, explain the CISD process, and set some basic ground rules for the debriefing. These normally include the follwing types of requests:

   - Only people involved in the incident take part in the debriefing

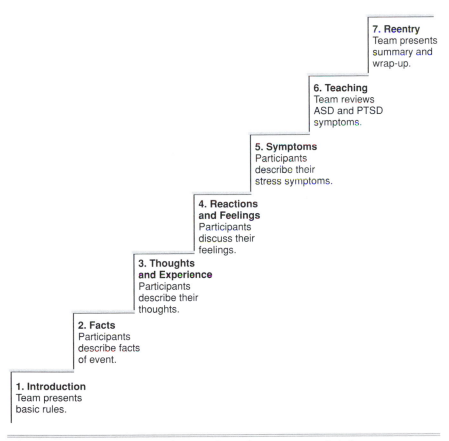

Figure 4.4 *Seven Steps of Critical Incident Stress Debriefing*

- The proceedings are confidential (participants may be asked to make a confidentiality pact)
- Participants are off-duty and can remain until the debriefing ends
- CISD is not a critique of the operation

2. *Facts*. Participants are asked to describe what happened at the event in a factual manner. Each participant is given an opportunity to talk about his or her role, but is not required to do so.

3. *Thoughts and Experience*. Participants are asked to describe some of the thoughts they had while involved in the incident. Doing so helps participants to personalize the event and move from facts to individual experiences.

4. *Reactions and Feelings*. In this step, participants are asked what the worst part of the event was for them and why they were bothered by it. The controlled group discussion allows people to vent their thoughts and feelings about the event.

5. *Symptoms*. Group members are asked to describe stress symptoms experienced from the beginning of the incident until the present.

6. *Teaching*. CISD team members provide information for the participants on normal as well as abnormal stress responses. They review signs and symptoms of acute stress disorder and post-traumatic stress disorder. Useful information for reducing stress symptoms, the advantages of communicating with friends and family, as well as the benefits and self-benefits of helping each other are also presented.

7. *Reentry*. This step includes a summary and wrap-up. The team leader summarizes the debriefing and provides participants with the opportunity to ask questions before they leave. Participants may wish to formulate specific plans that help them gain closure on the stressful event, such as organizing a memorial fund or attending the funeral of a fallen colleague. Team members also provide information on mental health and counseling services for individuals needing additional assistance.

Availability of CISD

Since its inception in the 1980s, CISD has become widely available. Most departments today have some mechanism in place to offer CISD for their members who are affected by critical incidents. The popularity of CISD has exposed significant numbers of emergency responders to the process. Many have participated in multiple CISD sessions. Familiarity with the CISD process has given many responders and emergency service managers the information and knowledge to conduct informal CISD sessions of their own. Today, it is likely that many shifts and crews debrief themselves following major incidents (see Figure 4.5). There is little to suggest that the outcomes of informal debriefings would be any different from those of formal CISD [5].

In light of growing evidence that suggests single-session debriefing may not be helpful, both the U.S. National Institute of Mental Health and the World Health Organization have recommended against it, offering instead a list of "best practices" supported by the most recent research evidence. These best practices constitute what is now referred to as "psychological first aid" and are recommended for people exposed to extremely stressful events:

1. Assess and provide for immediate physical needs (medical care, water, food).
2. Ensure physical safety (housing, shelter, clothing).
3. Offer practical help (babysitting, insurance paperwork, phone calls).
4. Help contact normal sources of support (friends, family, spiritual community).
5. Facilitate contact with loved ones (local and distant).
6. Educate about normal responses to extreme stress (ASD and PTSD).
7. Support real-life task decisions (help prioritize tasks that need attention).

Further recommendations include limiting any discussion of the stressful event during the first two to three weeks to only what the individual wants to talk about. Mental health profes-

Figure 4.5 *Critical Incident Stress Debriefing (CISD) as Means to Relieve Stress (Source: Reproduced with permission from First Responder: Your First Response in Emergency Care, American Academy of Ortho-paedic Surgeons and National Safety Council®, p. 25, Figure 2.4. Courtesy of Jones and Bartlett Publishers.)*

sionals are advised that continually retelling a traumatic story in the early weeks after an event may be unhelpful and may increase the risk for persistent PTSD. Therapists are also cautioned not to overwhelm affected individuals with information [9, 10]. These recommendations are valuable not only to mental health workers, but to emergency service leaders as a "best practice" for assisting members involved in critical incidents today.

CISD and other debriefing programs like it may have swung the pendulum in the opposite direction of the John Wayne syndrome. Once perceived as invincible and having emotions under control and nerves of steel, emergency responders today are sometimes portrayed as overly sensitive and in need of emotional assistance with virtually any critical incident. This new stereotype may be just as harmful as the John Wayne image of the past. In fact, studies have shown that emergency responders do have special coping skills and abilities to deal with the stressors of their jobs [8]. Perhaps a more accurate portrayal of today's emergency responder would be that of a person well prepared for any incident, but susceptible at times to the effects of extraordinary stress.

SUMMARY

Emergency responders have multiple opportunities to see and experience events that others might never encounter in their lifetime. Responders have coping skills that allow them to deal effectively with stressful calls, but unusual events can produce stress symptoms. Acute stress disorder (ASD) is the label given to symptoms lasting less than four weeks and post-traumatic stress disorder (PTSD) applies to symptoms lasting longer than four weeks. No amount of experience can prevent stress symptoms, but experience does allow responders to anticipate their reactions and learn how long they typically last before they feel normal again. When a responder's life is disrupted by stress symptoms for more than two to three weeks, he or she should seek counseling assistance.

For years, emergency responders were portrayed as callous and unaffected by emotion. This John Wayne syndrome led many responders to believe that emotional effects of extremely stressful calls were signs of weakness. Suppressing stress symptoms is extremely unhealthy and can lead to a myriad of problem behaviors as well as resignation or termination from the emergency services. Today, responders are recognized as human beings with genuine emotions and the capability to be personally affected at times by critical incidents.

Critical incident stress debriefing (CISD) is a formalized session run by mental health professionals and peer support personnel for groups of responders who have experienced a highly charged event that tests their coping capabilities. Since the 1980s, CISD has sought to reduce the impact of stressful events for emergency responders and swiftly return them to their normal routines. Widespread exposure to CISD has placed debriefing skills into the hands of many responders and their department leaders who often debrief informally without the aid of CISD teams. Recent mental health suggestions on "psychological first aid" for people exposed to extreme stressors provide expert guidance to emergency service departments on assisting members involved in critical incidents.

REFERENCES

1. American Psychiatric Association, *Diagnostic and Statistical Manual of Mental Disorders,* 4th ed., Washington, DC, 2000.
2. Wagner, Dieter; Heinrichs, Markus; and Ehlert, Ulrike, "Prevalence of Symptoms of Posttraumatic Stress Disorder in German Professional Firefighters," *American Journal of Psychiatry,* Vol. 155, No.12 (1998): 1727–1732.
3. National Institute of Mental Health, *Facts About Anxiety Disorders,* Washington, DC: Public Health Service, October 2001.
4. Mitchell, Jeffrey T., "When Disaster Strikes . . . the Critical Incident Stress Debriefing," *Journal of Emergency Medical Services,* Vol. 8 (1983): 36–39.
5. van Emmerik, Arnold A. P.; Kamphuis, Jan H.; Hulsbosch, Alexander M.; and Emmelkamp, Paul M. G., "Single Session Debriefing after Psychological Trauma: A Meta-analysis," *The Lancet,* Vol. 360 (2002): 766–771.
6. Bledsoe, Byran E. "EMS Myth #3: Critical Incident Stress Management (CISM) Is Effective in Managing EMS-Related Stress," *Emergency Medical Services,* Vol. 32 (2003): 77–80.
7. Mitchell, Jeffrey T., "Development and Functions of a Critical Incident Stress Debriefing Team," *Journal of Emergency Medical Services,* Vol. 13 (1988): 42–46.
8. Moran, Carmen C., "Individual Differences and Debriefing Effectiveness," *Australasian Journal of Disaster and Trauma Studies,* 1998-1.
   www.massey.ac.nz/~trauma/issues/1998-1/ moran1.htm
9. National Institute of Mental Health. "Mental Health and Mass Violence: Evidence-Based Early Psychological Intervention for Victims/Survivors of Mass Violence – A Workshop to Reach Consensus on Best Practices," NIH Publication No. 02-5138, Washington, DC: U.S. Government Printing Office, 2002.
   www.nimh.nih.gov/research/massviolence.pdf
10. World Health Organization. *Mental Health in Emergencies: Mental and Social Aspects of Populations Exposed to Extreme Stressors.* Geneva: World Health Organization.
    www5.who.intmental_health/download.cfm?id=0000000640

Stress at Home 5

Work in emergency services is often a thankless job. It can be tremendously challenging at times, and very boring at other times. Emergency responders have a crucial need for support of family and friends at home. Home is the emergency responder's life away from emergency services. Home is where emergency responders live, the significant relationships in their life, and the family and friends they depend on. Ideally, home is a refuge from the difficult job of being an emergency responder. At times, however, home can become an additional source of stress. When that happens, the difficulties typically involve communication problems.

Emergency responders are different from ordinary people. Their jobs are different, and, as a result, their families and relationships are different. This chapter is not about how to communicate; many workshops, tapes, and books are available for people who want to improve their communication skills. This chapter is, however, about keeping the emergency responder's home life happy. In order to do that, we'll take a close look at how emergency services profession influences the home life of the emergency responder. Remember that conflict never develops without the help of both parties. To that end, we'll take a look at both sides of emergency responder relationships with an eye to making the home a better place.

UNDERSTANDING THE SOURCE OF HOME LIFE PROBLEMS

Although the home represents their life away from work, separating home life from work life is not possible for

Key Point: *Although the home represents life away from work, separating home life from work life is not possible for emergency responders.*

emergency responders. Not all difficulties at home really begin there. Sometimes, troubles begin at work and spill over to the home. Other times, problems at home affect performance and satisfaction at work. It would be rare for any person having serious problems at work not to see those difficulties spill over into the home and vice versa.

When troubles arise at home, considering the source of the problem is important. Knowing the origin helps determine how and where to approach the problem. Properly fighting a fire you cannot see is difficult, if not impossible. Symptoms of stress, like smoke from a fire, can sometimes be found at considerable distances from the source of the problem.

Emergency responders have more difficulty with marriages, relationships, and raising children than do others in society.

Key Point: *Rates of marital and family problems among police, fire fighters, and medics far exceed national averages.*

Studies abound reporting rates of marital and family problems among fire fighters, medics, and police that far exceed national averages [1, 2]. In Chapter 2 we saw that the same personality traits that make an effective emergency responder can also increase conflict in the workplace and lead to lower self-esteem. These and other emergency responder personality characteristics lead to conflict at home as well.

The difference between work and home is often dramatic. Emergency responders may not be aware that their family members and significant others often do not share their adrenaline-driven personalities. Knowing the differences between an ordinary person in society and the roughly 10 percent who share the emergency responder's addiction to adrenaline can provide tremendous insight into behaviors of the people we live with. Likewise, spouses and significant others often have little awareness or understanding of exactly what makes an emergency responder tick. Learning about the unique personalities of emergency responders can be a big eye-opener for family and loved ones.

ISSUES AFFECTING HOME LIFE

Aside from personality differences, the role of an emergency responder itself has considerable impact on his or her relationships and life at home. From the perspective of the families of emergency responders, five issues related to emergency services work stand out as different from all other lines of work: work schedule, loyalty, risk of danger, stressors carried home, and public opinion.

Work Schedule

Emergency services never close. The job demands that workers be available 24 hours a day, 7 days a week, 365 days a year. Emergency responders must work nights, weekends, and holidays. They miss family events, birthday parties, scout meetings, school plays, and Little League games because of their erratic work schedules (see Figure 5.1). Parties, evening social events,

Figure 5.1 *Little League Game (Source: Courtesy of Cody Langway, Middleboro All Stars)*

weekend outings, and sporting events can conflict with work and on-call schedules. Longer than usual work shifts can take an emergency responder away from home for what might seem like forever to his or her family.

When children are involved, the emergency responder's spouse or partner may at times feel like a single parent. The added burden of trying to assure that the emergency responder is able to sleep without interruption can add to this already heavy weight. Children themselves may feel they come from a single-parent family, interpreting the absence of their emergency responder parent as a lack of concern for them. On-call status adds to work schedule issues by introducing uncertainty into an already less-than-normal presence at family events. Nearly every study conducted on stress in emergency services families highlights work schedules as the most significant issue affecting emergency responders' home lives.

Key Point: Nearly every study conducted on stress in emergency services families highlights work schedules as the most significant issue affecting emergency responders' home lives.

Loyalty

Emergency responders characteristically demonstrate unwavering loyalty to each other, their leaders, and their departments. Their drive to finish what they start can cause them to remain at emergency scenes long after their work shifts are over. Their enthusiasm and commitment to emergency services may lead them to work longer and more frequent shifts. More time spent on the job means less time at home, and that becomes an additional source of stress. Responders may tell their loved ones that home and family come first, but their long hours and enthusiasm for their jobs may communicate the opposite.

Key Point: Responders may tell their loved ones that home and family come first, but their long hours and enthusiasm for their jobs may communicate the opposite.

Additional training the emergency responder takes to improve performance or gain a promotion also means more time away from home. Study for exams detracts from time spent with loved ones. When questioning the emergency responders' true priorities, family and friends cannot help but notice the unusually strong camaraderie that emergency responders share with one another. Family members may become jealous of these attachments and wonder whether emergency services or family has the greater share of the emergency responder's loyalty [3]. They also have difficulty understanding the emergency responder's motivation to "drop everything" and respond to an emergency and may wish that their needs and interests got similar responses.

Risk of Danger

Danger has always been a concern to the family and friends of emergency responders, probably more so than is ever expressed. Since the terrorist attacks of September 11, 2001, concern for danger has never been more present in the minds of those who live with and care about emergency responders. Workers in the general population of the United States have a national work-related accident fatality rate of 5.0 deaths a year per hundred thousand workers. The yearly fatality numbers for emergency responders far exceed national averages at 16.5 for fire fighters, 14.2 for the police, and 12.7 for emergency medical service workers [4]. Whereas emergency responders condition themselves to ignore the dangers they routinely face, their families and loved ones have trouble doing the same. The ever-present reality is that each time an emergency responder leaves home, he or she may never return.

> *Key Point: Danger has always been a concern to the family and friends of emergency responders.*

Compounding this anxiety is the extremely unpredictable nature of danger in emergency services. Even the most routine call can turn into a dangerous situation. Danger can extend from work into the home as well, particularly for law enforcement emergency responders. Weapons brought home or kept in the house pose a constant danger to curious children. Threats to public safety and law enforcement personnel may follow them from work to home and may at times include their families and loved ones. All of these concerns increase the potential for troubles at home.

Stress Carried Home

Whether spoken or unspoken, disturbing occurrences that take place on the job can't be left behind when the emergency responder returns home (see Figure 5.2). Families know this, and even small children seem to know when a parent or adult is upset or feeling anxious. Traditionally, emergency responders do not take their work home with them, trying to protect or shelter their families from the difficult situations seen on the job.

Whereas workers in no other occupation seem more capable of controlling their emotions and feelings than do emergency responders, their attempt to use the same tough-guy shell at home that they use at work leads to problems. Beyond the consequences to the emergency responders of their bottling up negative experiences, their silence leaves families and loved ones to draw their own conclusions about what might be affecting the emergency

Figure 5.2 *Emergency Responders at Accident Scene (Source: Jim Ruymen / Reuters)*

responder. Not surprisingly, without being given clues to conclude otherwise, families often decide that *they* are the problem.

Isolation that ensues from lack of understanding causes tension for everyone
involved and can destroy relationships
and break down a family unit. Negativity
acquired from dealing with some of the
worst elements of society can accompany
emergency responders into their personal lives. This cynicism and down-
on-the-world view can be distressing to
those at home.

> **Key Point:** *Isolation that*
> *ensues from lack of under-*
> *standing of the emergency*
> *responder's bottling up*
> *negative experiences causes*
> *tension for everyone involved*
> *and can destroy relationships*
> *and break down a family unit.*

Public Opinion

Emergency service providers are at the forefront of public and
media scrutiny. Good or bad, their performance is continually
on stage for the world to view. Having an emergency responder
in the family is a source of pride, added prestige, and status.
Public opinion can, however, turn against the emergency responder. Allegations of incompetence, beatings, oppressive behavior, or theft leveled against a single member of a police, fire,
or emergency medical service department can cause the spread
of negative public opinion toward all responders. The continu-
ous seesawing public perception of the emergency responder as
either an invaluable asset or a dispensable resource is trying,
not only for responders but also for their families and significant others. In addition, the firsthand knowledge family members have of the intense loyalty, effort, and dedication their
emergency responder has for his or her work makes the public
criticism all the more stressful and offensive.

COMMUNICATION AS THE KEY FACTOR

Value of Feedback

One word summarizes the effect of emergency services work on
relationships at home: *understanding*. Emergency responders
believe that their loved ones don't understand them or their
emergency service work; loved ones believe that emergency responders don't understand their needs and interests. In the

Key Point: *Without feedback and good communications, emergency responders' significant others will believe that something they did caused the emergency responder's behavior toward them.*

field of relationship counseling, many problems between couples can be traced to a lack of feedback. For emergency responders, who can have very different behaviors from mainstream society, the need for feedback takes on even greater significance. Without feedback and good communications, emergency responders' significant others will believe that something they did caused the emergency responder's behavior toward them.

Communication is not always spoken. Consider the following cases:

> **Case:** At the end of long and tiring shift, an exhausted fire fighter returns home where he prefers to read the paper and take a nap than spend time with his family. He feels physically tired and emotionally drained. Rather than recognizing his exhaustion as the source of his behavior, his wife and family interpret his lack of interest in them as a message of rejection.

> **Case:** A police officer is unable to change his work schedule in order to attend his son's Little League games Tuesday afternoons. His son interprets the absence as a message that his father doesn't care about him.

> **Case:** A paramedic experiences a particularly gruesome call involving the death of two young children. On arriving home, she immediately contacts a good friend from her paramedic class and makes arrangements to meet at a local pub that evening. Her husband interprets this as a message that he is not good enough to lend adequate support to his wife in difficult times.

These scenarios may well be examples of misinterpretation, but miscommunication occurred nonetheless. Spoken words would have served in each situation to clarify or perhaps to contradict the unspoken message. The value of feedback between individuals involved in relationships can never be emphasized enough.

Effect of Responder Personality Characteristics on Home Life

In earlier chapters we looked at personality characteristics unique to emergency responders and how these traits helped make them more effective on the job. We also saw how these same characteristics can hinder work relationships and increase job stress. Not surprisingly, there are two characteristics found in emergency responders that interfere with relationships at home and lead to family troubles: high expectations and action orientation.

High Expectations

The high expectations that emergency responders hold for themselves and their coworkers often extend to their families as well. This can have particularly destructive effects on relationships. Allowing others, especially children, to make mistakes and learn from them is critical to the development of self-confidence. Not allowing room for failure or mistakes can have a destructive effect on family dynamics.

Action Orientation

Action orientation leads emergency responders to continually seek activities that fulfill their need for stimulation and prevent boredom. Their tendency to pursue action-oriented activities outside of work attracts them to acquire boats, guns, motorcycles, and other, frequently costly, "toys." The financial burden associated with these purchases and expenditures and the added time away from family who may not be invited or not interested in the thrill-seeking recreational pursuits of the emergency responder often has a stressful impact on family life.

Communication Skills for the Emergency Responder

Just knowing the potential sources of conflict on the home front is not enough for the emergency responder to avoid them. Because conflict occurs with the help of both parties, improving communication and creating a better relationship requires effort from both sides. We've already seen that communication is more than just talking, it's behavior as well. That's an important point to keep in mind. Evaluating our behaviors for the messages we might be sending can help prevent problems before they get out of hand.

What specifically does the emergency responder need to know in order to have the best possible home life, and what specifically do their significant others need to know for their happiness? We've already said that communicating is essential to a happy home life. Everything an emergency responder can learn about communication skills pays dividends. Communication skills for emergency responders center on sharing content, listening, and spending time.

Sharing Content

Content is what you share with those you live with and those you love. The tough shell used on the emergency services job is not intended for use at home, but many emergency responders have trouble leaving their emotional armor at the door when they return home from work. Doing this requires thought and effort. Responders may feel a need to protect others from gruesome occurrences they encounter at work by keeping these experiences inside. They may also be afraid that others would consider them troubled or strange for being able to tolerate horrific events on the job. Good and strong relationships are built on self-disclosure and shared experiences. See, for example, the following self-disclosure case:

> **Case:** "Would you like to take a trip in a hot air balloon at the outdoor festival next weekend?" Bonnie asked Dave, her police officer husband of 3 years. "Have I ever told you that I'm scared of heights?" Dave replied. She thought about it for a moment. "No, but now that you bring it up, it makes perfect sense." Bonnie remembered the many times Dave had backed out of trips or vetoed her ideas for outings. It was like a sudden revelation: Every instance had involved great heights. There were mountains, gondolas, airplanes, and parachutes. "I wish I had known this sooner," Bonnie said, "I was starting to think it was me you had an aversion to."

Key Point: Good and strong relationships are built on self-disclosure and shared experiences.

If emergency responders want meaningful relationships with people outside of work, they need to talk about themselves and their feelings. They need to share their reactions to events in their life and not be afraid to let others know how important they are to the emergency responder or how much the responder needs them in his or her life. Remember: Silence sends a message. Refusing to talk or keeping quiet and bottling up thoughts inside yourself sends the clear message that "I'm not interested in sharing with you." Sharing private thoughts and feelings with others is risky; it makes you vulnerable. Emergency responders are control-oriented people who can be very uncomfortable revealing their inner thoughts and weaknesses to others. If that characteristic fits you, recognize it and work to overcome it. Your home life will benefit tremendously.

> **On the Lookout:** *Remember that silence sends a message.*

> **On the Lookout:** *If you are uncomfortable revealing your inner thoughts and weakness to others, recognize this characteristic and work to overcome it.*

Listening

Emergency responders can be superb listeners, when they want to be. They can pick out details of a motor vehicle crash from a police radio in the next room while they read, watch television, and carry on a conversation all at the same time. They frequently need to make very quick assessments at emergency scenes, collecting spoken, visual, and physical clues to rapidly decipher the message and act accordingly (see Figure 5.3). Quickly sizing up situations and people requires judgments—and usually involves some prejudgment.

Emergency responders tend to carry this form of listening with them to other places. The responder may listen to the first sentence or two from his or her spouse or child and rapidly conclude what the message must be. Sometimes this works—but often the real message comes after the emergency responder has stopped listening. Knowing that emergency responders have a tendency to prejudge those they listen to can help them to improve their listening skills. Suspend judgment, listen without prejudice, perhaps even paraphrase what

> **On the Lookout:** *Suspend judgment, listen without prejudging, perhaps even paraphrase what others tell you so that both of you are certain that you understood what you heard.*

Figure 5.3 *State Trooper Interviewing Witnesses at Accident Scene (Source: © Syracuse Newspapers/Gary Walts/The Image Works)*

others tell you so that both of you are certain that you understood what you heard. The people who emergency responders love and depend on for support need to be understood, and emergency responders need their loved ones to understand them. But listening is not the only thing involved in understanding—you also have to spend time together.

Spending Time

Time is an enemy of good communication and of healthy relationships. Listening takes time. Sharing experiences takes time. Relationships take time. But emergency responders have only a limited amount of time in their lives. Bizarre work schedules, multiple jobs, continual training, and the need for rest can leave little time for family and friends. Ultimately, relationships disintegrate when the people involved lose their commitment to spending time together.

If emergency responders really want to keep their relationships healthy and happy, they need to commit to spending time together. Even if that time is only 30 minutes per week, the idea of reserving planned, purposeful time together sends an important message. It tells others that they are important to you and that, no matter how busy your life becomes, you continue to care about them. Rather than just thinking about spending some time with those you care about, make plans to do so, and carry out those plans. In addition, make a commitment to yourself to do it regularly.

> **On the Lookout:** *If you really want to keep your relationships healthy and happy, you need to commit to spending time with your loved ones.*

> **On the Lookout:** *Rather than just thinking about spending some time with those you care about, make plans to do so, and carry out those plans. In addition, make a commitment to yourself to do it regularly.*

Awareness

Awareness is important as well. If family members are aware of the unique characteristics of emergency responders, they'll have a better understanding of their motivations and behaviors. Emergency responders who are educated to recognize tendencies in themselves toward high expectations and thrill-seeking recreational activities can monitor themselves to avoid related conflicts. Departments can assist families by providing information and education on the unique aspects of emergency services work. Supplying family or significant others with such information can provide them with valuable insights about their emergency responder.

Communication Skills for Significant Others

Communication skills for the significant others of emergency responders center on showing interest, listening, and expressing confidence.

Conveying Interest

Emergency responders are skilled at controlling their emotions. That ability allows them to be calm and collected even during the most chaotic of crises. Oftentimes, they feel compelled to protect others from exposure to the difficult circumstances they encounter by keeping these buried inside themselves. People who live with and care about emergency responders can lend encouragement by expressing interest in them. Conveying interest or curiosity in the emergency responder's work in general and his or her latest shift in particular opens the door to communication. Voicing encouragement and showing interest are particularly helpful in relationships with emergency responders because emergency responders tend to shy away from sharing feelings that may make them vulnerable. See, for example, the following "conveying interest" case:

> **Key Point:** *Conveying interest or curiosity in the emergency responder's work generally and his or her latest shift in particular opens the door to communication.*

Case: As he unlocked the front door of his apartment, John knew he would probably never forget his last EMS call of the shift: a young father gunned down in a liquor store robbery, in plain view of his two little children. In an effort that bordered on superhuman, John and his partner did everything they could to keep the young man alive and it seemed like he stood a fighting chance. Minutes from the emergency department, the battle took a turn for the worst, and ultimately nothing could keep the man from dying. The scene was gruesome; the ambulance splattered with blood, and the hospital another ghastly sight. These were not the sort of stories John took home with him; it seemed bad enough that he was there. Why would he want his wife, Lisa, to be burdened with the ills of society?

Yet, Lisa was interested. She often asked John about his work and how his hours were spent. Today was no exception. "You look frazzled," Lisa said, greeting John at the door, "How was your shift?" John paused for a moment, and then decided to tell his wife about the horror he'd just experienced. Leaving out the gore and blood, he explained the sadness and powerlessness he felt, revealing as much as he felt comfortable telling his wife. To his surprise, Lisa did not seem shocked or horrified. Instead she was sympathetic and thankful that John had shared with her something of himself. Perhaps this was something he should do more often, John thought to himself.

Listening

Listening is an issue not only for emergency responders who tend to jump to conclusions about what they hear from others, but for their significant others as well. Emergency responders make judgment calls every day. Their role requires them to do so. Their job requires them to be on target constantly, and they are accustomed to a high degree of success resulting from their decisions. They commonly tell others what to do in difficult situations and are geared toward needing little if any direction themselves. As a result, they are people who are very sensitive to criticism and who may not welcome guidance from others in their decision making. An awareness of this sensitivity can be helpful for significant others in listening to and communicating with emergency responders. Avoiding judgmental or critical comments will facilitate longer and more meaningful conversations between emergency responders and those they love.

Expressing Confidence

Expressing confidence in the emergency responder's worth and his or her importance to the significant other and the family is the third element of communication to keep in mind. People involved in emergency services have extremely high expectations of themselves and of others. They need to work hard not to go overboard in their expectations of themselves, their coworkers, their friends, or their family, and doing so is at times a challenge for them. The high expectations they apply to themselves exact a cost in self-esteem, although that might seem hard to believe, given all the accomplishments of the typical emergency responder. Expressing confidence in emergency responders lets them know that it is safe to talk and share details about themselves. Despite what they do for their community, they get few compliments. Those who express confidence in them are their greatest allies.

Key Point: *Expressing confidence in emergency responders lets them know that it is safe to talk and share details about themselves.*

ADDRESSING BEHAVIOR PROBLEMS

Finally, home life is never perfect. Problems will always occur. When they happen, they need to be dealt with. Behaviors not addressed will never change. When emergency responders have or cause difficulties at home, they need to address the problems directly. Remember also, that problems—not people—need to be attacked. When the time is right for sitting down and talking about an issue at home, you might use the following approach to address the problem:

- Say what's bothering you.
- Say how it makes you feel.
- Suggest a solution.
- Invite comment.

Case: Rick is a 35-year-old fire fighter whose new wife, Maria, calls him almost hourly when he's at work. He's not happy with the frequency of these calls. Using the problem-solving approach just described, Rick could approach his wife at a comfortable time, when they are both able to talk, to convey the following message to his wife:

- "Calling me every hour while I'm at work bothers me."
- "I feel distracted, nervous, and unable to focus on my job."
- "I'd like to suggest that I call you twice each day."
- "Of course, you could always call me in an emergency."
- "How does that sound to you?"

Certainly there are many ways to discuss problems between people, and no single method will be comfortable for everyone. The critical point for everyone involved in a relationship is to make sure that problems are addressed in some fashion as soon as practically possible. Small things ignored now become big problems later.

SUMMARY

For the emergency responder, home should be a retreat from the demanding job of providing emergency services. When home, family, or friends seem more nerve-racking than support-ive, communication problems are the likely culprit. Knowing whether troubles originated at work or at home is the first step in dealing with them. Problems not addressed grow bigger and eventually spill over to both work and home.

Awareness of unique personality characteristics of emergency responders is important for both responders and their significant others. This knowledge provides a better understanding of each other's behaviors. Knowing trigger areas that typically disrupt relationships helps responders and their families to address small problems before they snowball into big issues—and lead to isolation. Aberrant work schedules, unwavering loyalty to emergency services, constant risk of danger, stressors carried home, and offensive public opinions all have a significant impact on relationships in emergency responders' families and can lead to troubles.

Improved communications means feedback between individuals to assure that messages are understood correctly. Emergency responders should pay particular attention to leaving their emotional armor at the door after work. Good relationships at home depend on self-disclosure and shared experiences. Significant others can encourage better communication by expressing interest in their emergency responder's job, and by encouraging the growth of a trusting, open environment.

When problems occur, they must be addressed or change will never occur. No relationship is without problems, except for those that are over.

REFERENCES

1. Roberts, Nicole A., and Levenson, Robert W., "The Remains of the Workday: Impact of Job Stress and Exhaustion on Marital Interaction in Police Couples," *Journal of Marriage and Family* 63, No. 4 (November 2001): 1052–1067.
2. NYC Patrolmen's Benevolent Association. *Program for the Reduction of Stress for New York City Police Officers and Their Families—Final Report*. U.S. Department of Justice Document # 185845. www.ncjrs.org/pdffiles1/nij/grants/185845.pdf (accessed December 12, 2000).
3. Shelton, Ray, and Kelly, Jack, *EMS Stress: An Emergency Responder's Handbook for Living Well*, Carlsbad, CA: JEMS Communications, 1995.

4. Maguire, Brian J.; Hunting, Katherine L.; Smith, Gordon S.; and Levick, Nadine R., "Occupational Fatalities in Emergency Services: A Hidden Crisis," *Annals of Emergency Medicine* 40, No. 6 (2002): 625–632.

Shift Work and Sleep   6

The major focus of the previous chapters has been on stress involving individuals—that is, emergency responders and the many people in their lives. Our interactions with others and the many ways we respond to the demands others place on us comprise the most significant sources of stress in our lives [1]. But not all stress involves people. Our environment can be equally stressful, and sometimes more so than the people in our lives. This chapter looks at an aspect of the environment that has significant impact on emergency responders: shift work. Shift work impacts the emergency responder's ability to sleep and feel rested.

SHIFT WORK AND SLEEP DEPRIVATION

Emergencies happen at all times of the day and night. For the emergency responder, therefore, shift work is a fact of life. Unfortunately, few emergency responders are ever taught how to adjust their lives to changing shifts. Most struggle to cope with shift work through trial and error, often working to the point of exhaustion. Some die trying to cope. Nearly one-fifth of the American work force is employed outside of the standard daytime work hours of 6 am to 6 pm [2]. Their shifts vary as much as their jobs: from manufacturing to engineering to emergency response. What shift workers have in common is sleep deprivation— that is, not getting enough sleep. Could each of the circumstances in the following cases have occurred because of sleep deprivation?

Case: Joe and his partner Sam are both veteran para-medics. Neither of the two had any sleep since they started work at 7:00 pm. At 5:00 am they were dispatched to their ninth call of the shift. Their patient was an elderly woman with severe difficulty in breathing due to acute heart failure. After several tries, neither medic was able to start an IV and their patient deteriorated into respiratory arrest. They also had difficulty with the airway and were unable to intubate. The patient died in the emergency department.

Case: At 3:00 am a burglary in progress was transmitted by Mobile Data Terminal to two police patrol units. Car 632 did not acknowledge the MDT, nor did she answer when the police dispatcher called her unit repeatedly on the radio. The shift supervisor later found the officer sleeping soundly in her patrol unit parked behind a local diner.

Case: At 1:00 am, Engine 69 was dispatched to a report of a carbon monoxide (CO) detector activation in a private residence. The engine officer asked Nate, an experienced fire fighter, to obtain CO readings outdoors and inside the residence. Despite having used the CO meter repeatedly in the past, Nate couldn't recall how to power the unit on.

For the average person, enough sleep means feeling rested and alert during waking hours. Although shift work in and of itself is not a cause of sleep deprivation, people who work while the majority of the world sleeps tend to develop unhealthy sleep habits. Disrupted sleep poses additional problems for many emergency responders. An understanding of sleep and what adjustments are necessary when working shifts can help emergency responders to feel more rested.

> **Key Point:** *What shift workers have in common is sleep deprivation—that is, not getting enough sleep.*

Night Shift Workers

Studies show that night shift workers experience increased stress. The divorce rate among night shift workers is higher and they are more likely to have accidents both at work and while traveling to and from work. Although the specific reasons are not clear, night workers also seem to have a higher frequency of heart disease, stomach and digestive disorders, and menstrual irregularities, yet their overall death rates are no higher than those of the general population [2].

People polled by the National Sleep Foundation in 2002 believed that lack of sleep might be responsible for both injuries and health problems. Of those surveyed, 85 percent also felt that inadequate sleep causes difficulty in getting along with others [3]. Another recent study demonstrated increased blood pressure in subjects following a single night of sleep deprivation [2]. Even moderate levels of sleep deprivation, such as getting 6 or fewer hours of sleep per night, can produce serious impairments in performance. Alarmingly, it has recently been shown that people are largely unaware of how impaired they become without sleep, which may explain why sleep deprivation is common in our society: People falsely believe they can adapt to it [4].

The level of impairment after 20 to 25 hours of continued wakefulness has been shown to be equivalent to a blood alcohol content (BAC) of 0.10 percent, which is considered legally intoxicated in most jurisdictions of the United States [5]. The degree of impairment coupled with seeming unawareness might explain why major incidents such as the chemical plant leak in Bhopal, India on December 3, 1984, the Chernobyl nuclear power accident on April 26, 1986, and the Exxon Valdez oil spill on March 24, 1989, all occurred during early-morning hours.

Sleep Requirement

The amount of sleep people need differs enormously. Experts often cite 7 to 9 hours of sleep as the amount needed by the average adult [6]. In truth, there is no "normal" requirement for sleep. The need for sleep varies tremendously from person to person. What does remain constant, however, is the amount of sleep that each individual needs to feel rested, awake, and alert [7]. Individual adults require nearly the same amount of sleep throughout their life, although older people

> *Key Point: There is no "normal" requirement for sleep. The amount of sleep that each individual needs to feel rested, awake, and alert, however, remains constant.*

tend to sleep more lightly and find their sleep less restful. The average of the daily hours of sleep you need to feel rested over the course of a week will nearly always fall within 30 minutes of the average number of hours you need any other week. Uninterrupted sleep of at least 4 to 5 hours is also necessary to feel rested on awakening [7].

For study purposes, sleep researchers categorize adults who sleep less than 5 hours daily without daytime symptoms as "short sleepers," and those who sleep more than 10 hours daily as "long sleepers." [2] There have been people studied who required only 3 hours of sleep per day as well as some who were unable to function without 10 or more hours of sleep daily [7]. Both extremes are rare. In 2003, younger Americans (aged 18 to 54) responding to the National Sleep Foundation poll reported getting an average of 6.7 hours of sleep each weeknight and 7.1 hours of sleep on weekends [8]. Keep in mind that this reported average probably does not represent all the sleep needed for the study participants to feel rested. In fact, 37 percent of those surveyed reported being so sleepy during the day that it interfered with their daily activities a few days each month or more [6].

Sleep debt is a term used to describe the effect of cumulative loss of sleep. The greater the amount of sleep lost, the larger your sleep debt grows. The larger the sleep debt, the more sleep is needed to restore the body and mind to normal levels of performance [5].

SLEEP-WAKE CYCLE

We know that sleep is a function of the brain, yet years of research have failed to lead us to understand the purpose of sleep [2]. It seems reasonable to assert that sleep restores our energy level, especially since a good sleep produces feelings of rejuvenation and energy.

Circadian Rhythm

Emergency responders faced with the challenge of feeling rested may find it useful to review the current knowledge about how the brain and body interact to produce our daily sleep-wake cycle. A small center in the base of the brain, the endogenous circadian pacemaker (ECP), maintains a daily rhythm in

our bodies that controls temperature, hormones, and wakefulness. This rhythm is referred to as our *circadian rhythm* and
runs on an average 24.5-hour cycle [2]. In
Figure 6.1, the upper curve illustrates
sleep drive buildup and the lower curve
illustrates circadian rhythm, indicating
wakefulness. The stronger the sleep drive
buildup, the more the upper curve pushes
down. The more wakeful the person, the
higher the circadian curve pushes up.
During each cycle, temperature, energy
level, digestion, and a host of other bodily
functions rise and fall in regular and predictable patterns. Since our normal 24-hour day is a full half
hour shorter than the average person's circadian rhythm, daily
adjustment is accomplished by light.

Key Point: *A center in the base of the brain maintains a daily rhythm in our bodies that controls temperature, hormones, and wakefulness. This rhythm, referred to as our circadian rhythm, runs on an average 24.5-hour cycle.*

Light

Light in the evening shifts the circadian rhythm later and light
in the morning shifts the circadian rhythm earlier, keeping us
on a sleep-wake schedule that matches our daily schedule.

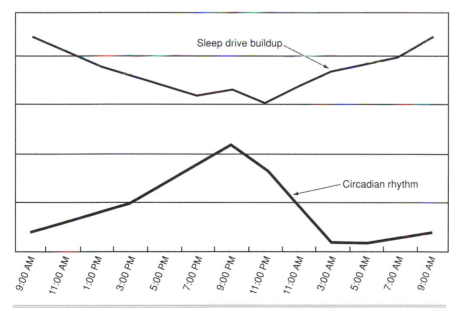

Figure 6.1 *Sleep Buildup Opposing Circadian Rhythm*

Other mechanisms are capable of shifting the circadian rhythm, but none comes close to the influence of light. Clearly, human beings are designed to sleep at night and be awake during daylight hours.

NREM and REM Sleep

The influences of circadian rhythms do more than allow the initiation of sleep and return to wakefulness. They are also closely associated with two distinct stages of sleep, non-rapid eye movement (NREM) and rapid eye movement (REM) sleep. NREM sleep is associated with low body temperature, slow metabolic rates, and stillness (see Figure 6.2). During REM sleep, which occupies the second half of a full sleep period, the body and brain are more active. Separating these stages of sleep—as might happen when a responder is called out in the middle of the night—can desynchronize their associated circadian rhythms.

Sleep Drive Buildup

The body, like the brain, also maintains a cyclical pattern that affect our daily sleep-wake cycle. This balancing (homeostatic) process of regulating sleep relies on a buildup of tiredness,

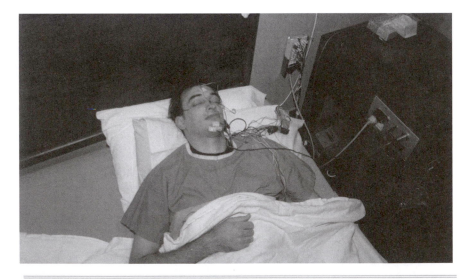

Figure 6.2 *Person Being Monitored in a Sleep Clinic (Source: Courtesy of the American Academy of Sleep Medicine)*

or sleep drive, that depends primarily on how long it has been since a person last slept. With a conventional sleep-wake schedule, sleep drive builds during the day and fades away with a night of sleep, leaving the individual alert the following morning. Under normal circumstances, the brain's circadian rhythm and the body's building sleep drive work in synchrony; therefore, as tiredness increases during the day, the circadian rhythm counteracts the propensity to sleep with increased hormonal, metabolic, and energy levels [2].

> **Key Point:** *The balancing process of regulating sleep relies on a buildup of tiredness, or sleep drive, that depends primarily on how long it has been since a person last slept.*

Conditions Affecting Sleep-Wake Cycle

Since the two systems must work synchronously to maintain our sleep-wake schedule, any interference or imbalance in either the circadian rhythm or the buildup of sleep drive can desynchronize the process and result in sleep-wake problems. Sleep researchers have identified a variety of physical, medical, and psychological conditions affecting the sleep-wake cycle. Of these, shift work and sleep deprivation are the two that most often trouble emergency responders [2, 5]. Aside from these issues, respiratory problems (such as sleep apnea), cardiovascular disorders, seizures, pain syndromes, psychiatric disorders, and a multitude of chronic medical conditions including diabetes, arthritis, kidney disorders, lung disease, stomach problems, cancers, and immune system disorders, such as HIV/AIDS, also affect the sleep-wake cycle.

The effect of alcohol in producing sleepiness and impaired driving ability provides a classic example of the interplay of circadian rhythm and sleep drive. Nearly twice the degree of sleepiness and driving impairment occurs when alcohol is consumed in the early afternoon hours compared to early evening consumption of an equivalent dose [2].

Overriding the Circadian System

For reasons that are poorly understood, some people seem to tolerate desynchronization of their circadian rhythm and sleep drive without much difficulty. These folks, as well as shift workers who seem to be able to reset their circadian clocks in response to varying work schedules, are clearly in the minority [2]. Sleep researchers have tried for many years to develop ways for shift workers to reset their circadian clocks. Efforts to accomplish this using forward-moving sleep-shifting techniques [5], bright light therapy [2], and other measures enjoyed varying degrees of success, but inevitably left shift workers again desynchronized on their days off when they resumed their normal schedules [2]. Currently, the majority of strategies endorsed for shift workers require overriding the circadian system and creating an artificial sleep drive. These same approaches are now recommended for travelers who experience sleep difficulties related to jet lag [2].

Shift Rotations

Few emergency responders have complete control over the shifts they work, but it may help to know that some shifts are easier to adjust to than others. Rotations that allow responders to remain on a shift for a minimum of three weeks reduce the need to continually reset workers' internal clocks and allow them to adjust to their schedules. When it is necessary to rotate shifts, research has shown that rotating shifts forward in a clockwise fashion results in greater job satisfaction, fewer accidents, and a healthier workforce compared to rotating shifts backward in time [5] (see Figure 6.3).

> **Key Point:** *Rotating shifts forward in a clockwise fashion results in greater job satisfaction, fewer accidents, and a healthier workforce compared to rotating shifts backward in time.*

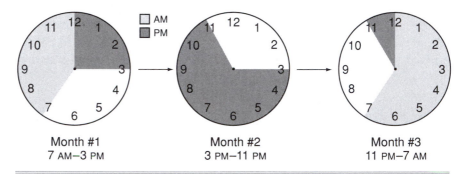

Month #1	Month #2	Month #3
7 AM–3 PM	3 PM–11 PM	11 PM–7 AM

Figure 6.3 *Shifts Rotating Forward*

"Larks" and "Owls"

Keep in mind that people are different: Some emergency re-sponders will adjust readily to changing shifts; others will have great difficulty working certain shifts. Knowing your own sleep-ing and waking preferences is important when you have a choice of shifts. Morning people, or "larks," who prefer to go to bed early and get up early, often adapt better to

> **On the Lookout:** *Knowing your own sleeping and waking preferences is important when you have a choice of shifts.*

day schedules. Evening and night types, or "owls," tend to ad-just better to night shifts. Other factors that influence prefer-ences are childcare concerns as well as reduced supervision and the slower tempo usually found on off-shifts [2]. Research has also shown that people who frequently have difficulty sleeping (i.e., insomniacs) often experience the greatest trouble adjust-ing to evening and nighttime work schedules [2, 5].

SLEEP HYGIENE

Sleep hygiene is a term given to promoting good-quality sleep of sufficient duration. (See Figure 6.4 for sleep hygiene tips.) Responders must decide for themselves the number of hours of sleep that they need in order to feel awake, alert, and well

> **Key Point:** *Sleep hygiene is a term given to promoting good-quality sleep that is of sufficient duration.*

rested. This requirement varies from person to person, but typi-cally remains constant for an individual adult over time.

Establishing Routines

Patterns are important to all sleepers, including shift workers. Try to establish a routine time for sleeping and stick to it. Workers who rotate shifts will probably find a few hours of time when they could always sleep regardless of whether they are on or off shift. Over-

> **On the Lookout:** *Try to establish a routine time for sleeping and stick to it.*

lapping sleep schedules to include these common times estab-lishes a solid routine and reduces change. An emergency re-sponder can stabilize his or her routine and lessen the impact

---

Sleep Hygiene Tips

Tips to Reduce Stress from Shift Work and Sleep Deprivation
❏ Decide how many hours of sleep *you* need to feel awake, alert, and rested.
❏ Establish a sleep routine and stick to it.

Tips to Induce Sleep
❏ Sleep in a dark room.
❏ Keep the room temperature cool (65–70° F or whatever temperature keeps your body neither too hot nor too cold).
❏ Assure quiet.
❏ Choose a comfortable bed.
❏ Avoid large meals before sleep, but don't go to bed hungry.
❏ Avoid caffeine (coffee, tea, cola, chocolate) and nicotine before bedtime.
❏ Avoid alcohol (which causes drowsiness but results in poor quality sleep).
❏ Consider use of a sleep aid (for occasional problems getting to sleep).
❏ See your health care provider to check for health problems that may be interfering with sleep (for frequent problems getting to sleep).
❏ Be aware that foods such as turkey, warm milk, and bananas help induce sleep.

Tips for Staying Awake on the Job
❏ Find something stimulating to do. (Activity promotes wakefulness.)
❏ Walk, run, or stretch. (Muscle activity is stimulating.)
❏ Brighten the lighting. (Light stimulates the brain to stay awake).
❏ Lower the temperature. (Cool, dry air promotes wakefulness; warm environment encourages sleep.)
❏ Go somewhere noisy. (Noise encourages wakefulness.)
❏ Stimulate your sense of smell with peppermint. (Peppermint increases alertness.)
❏ Consider using alerting medications (coffee, tea, cola, chocolate).
❏ Take a short nap (no more than 20–30 minutes) or go home, if you are overly tired.

Figure 6.4 *Tips for Achieving Sleep Hygiene (Source: Adapted from the National Sleep Foundation Web site at www.sleepfoundation.org)*

of shift work by overlapping sleep times between workdays and off days.

Naps

In between longer periods of sleep, naps have been shown to be extremely beneficial in restoring alertness [2]. Short naps of 20 to 30 minutes are preferred to longer naps of over 30 minutes. Longer naps can interfere with main sleep periods as well as initiate a "sleep inertia" that leaves the worker feeling foggy on awakening and can impair job performance [2]. A nap of several hours' duration is a very sound practice before starting a rotation to the

> *Key Point: In between longer periods of sleep, naps (preferably short naps of 20 to 30 minutes) have been shown to be extremely beneficial in restoring alertness.*

night shift or when working a sporadic night shift. A second nap after returning home will help the responder to recuperate enough energy to return to a normal routine [2, 5].

> **On the Lookout:** *A nap of several hours' duration is a very sound practice before starting a rotation to the night shift or when working a sporadic night shift.*

Environmental Factors

Achieving quality sleep requires attention to the environment.

Light

Light is the primary cue for the circadian rhythm cycle and bright lights have been used with varying degrees of success to improve night workers' shift adaptation. More importantly, responders need to assure that the environment in which they sleep is as dark as possible because light stimulates wakefulness. A sleeping mask or light-blocking curtains or shades can be used to further reduce daylight levels.

> **On the Lookout:** *Make sure that the environment in which you sleep is as dark as possible, because light stimulates wakefulness.*

Temperature

Temperature is a second factor intertwined with circadian rhythms. Maintaining a cool bedroom temperature—between 65° and 70°F—promotes optimal sleep for most adults.

> **On the Lookout:** *Maintaining a cool bedroom temperature—between 65° and 70°F—promotes optimal sleep.*

Noise

Interruptions pose significant problems for daytime sleepers. Choosing a bedroom away from extraneous noise sources such as bathrooms and outside traffic is helpful. Turning off phone ringers, beeping watches, cuckoo clocks, and other noise sources may or may not be possible, but is desirable. When noise persists, earplugs or "white noise" generated by an air conditioner, fan, or electronic device can be used to block the sounds. Posting a sign on your bedroom door will remind others that you are sleeping and prevent interruptions. Don't discount the benefits of a comfortable mattress as well.

Other Factors

Diet

Food is another factor that affects shift workers and sleep. Since circadian rhythms slow the digestive system at night, it is wise to avoid large meals and spicy or hard-to-digest foods during the night. These foods may cause nausea or indigestion. Likewise, eating large meals within 3 to 4 hours of sleeping can induce wakefulness or cause stomach discomfort, interfering with sleep quality [2]. Some foods have been shown to induce sleep and should be avoided while at work. These include turkey, warm milk, and bananas.

> **On the Lookout:** *Avoid large meals and spicy or hard-to-digest foods during the night.*

Exercise

Physical exercise has been shown to improve sleep quality [2, 5]. Exercise can induce tiredness and results in an improved overall sense of well-being. People respond differently to the biochemical changes induced by exercise, but, in general, it is wise to avoid intense physical activity for several hours before sleeping [2].

> **On the Lookout:** *Avoid intense physical activity for several hours before sleeping.*

Sleep-Related Medications

Alerting Medications

At times shift workers are tired despite their best efforts to practice good sleep hygiene, to watch their diets, and to exercise regularly. In these instances alerting medications may help to enhance wakefulness at work. The most widely used alerting medication is caffeine. Coffee, cola, tea, and even nicotine have stimulant effects on the body and can temporarily override both circadian rhythms and sleep drive [2].

> **Key Point:** *Coffee, cola, tea, and even nicotine have stimulant effects on the body and can temporarily override both circadian rhythms and sleep drive.*

Sleeping Medications

Likewise, obtaining adequate daytime sleep can be a significant problem for many night workers, despite their best efforts.

Intermittent use of sleeping medications can be helpful in some instances, and as long as they are used only occasionally, their risk is low [2]. Melatonin is an over-the-counter nutritional supplement available in the United States with demonstrated ability to promote sleep. Melatonin produces an effect on the circadian system similar to darkness and, in several studies, has been shown to improve sleep for night workers at doses of between 0.5 and 3 milligrams. In the United States, melatonin is classified as a nutritional supplement and not a drug; no serious side effects regarding its use have been reported to date [2]. Foods such as turkey, warm milk, and bananas may help to induce sleep as well. Alcohol is not a good sleep aid because it induces drowsiness but lasts for only a few hours, resulting in very poor-quality sleep [9].

> **Key Point:** *Melatonin is an over-the-counter nutritional supplement available in the United States with demonstrated ability to promote sleep.*

Sleep and Home Life

As pointed out in Chapter 5, work schedules involving rotating shift work is the most significant issue affecting an emergency responder's home life because it changes the schedule of every family member. Assuring quiet time for sleep in itself helps to produce stress in the family unit. Emergency responders are encouraged, therefore, to direct the same resourcefulness and creativity they apply to assuring adequate sleep for themselves to the larger issue of spending quality time with their family and loved ones.

> ***On the Lookout:*** *We encourage you to direct the same resourcefulness and creativity you apply to assuring adequate sleep for yourself to the larger issue of spending quality time with your family and loved ones.*

SUMMARY

Most stress involves interactions with the people in our lives, but shift work and sleep problems can lead to stress as well. Emergency responders can adjust their lives to reduce the stress of shift work and feel awake, alert, and rested each day. For the average person, enough sleep means feeling rested and awake during waking hours. There is no "normal" requirement for sleep, but the amount each individual needs to feel rested remains consistent over time.

Sleep is controlled by the regular rise and fall of brain circadian rhythms synchronized with a buildup of sleep drive in the body. Interference with circadian rhythms or sleep drive from any source, such as working night shifts, can result in sleep-wake problems and lead to sleep deprivation. Sleep hygiene refers to behaviors that promote good-quality sleep of sufficient duration. They are helpful for emergency responders to consider. Most sleep hygiene practices attempt to override circadian cycles rather than reset them. Diet and exercise habits are also effective both in promoting and interfering with good sleep. Alerting medications such as caffeine can enhance wakefulness at work. Intermittent use of sleeping medications or melatonin, a nutritional supplement, can improve sleep for night workers.

REFERENCES

1. Seaward, Brian Luke, *Managing Stress: Principles and Strategies for Health and Wellbeing,* 3rd ed., Sudbury, MA: Jones and Bartlett Publishers, 2002.
2. Lee-Chiong Jr., Teofilo; Sateia, Michael J.; and Carskadon, Mary A., *Sleep Medicine*, Philadelphia, PA: Hanley and Belfus, 2002.
3. National Sleep Foundation, *2002 Sleep in America Poll*. [online]. Washington, DC, 2003. www.sleepfoundation.org/. Accessed: April 15, 2003.
4. Van Dongen, Hans P. A.; Maislin, Greg; Mullington, Janet M.; and Dinges, David F., "The Cumulative Cost of Additional Wakefulness: Dose-Response Effects on Neurobehavioral Functions and Sleep Physiology from Chronic Sleep Restriction and Total Sleep Deprivation," *Sleep* 26, No. 2, (2003).
5. Lamond, Nicole, and Dawson, Drew, *Quantifying the Performance of Impairment Associated with Sustained Wakefulness*, The Centre for Sleep Research, Queen Elizabeth Hospital, South Australia. March/April 1998.
   [on-line] available: www.unionepiloti.it/diparti/tecnico/Documenti_dt/QuantifyPerf.pdf
6. National Sleep Foundation. *Myths and Facts About Sleep*. [online]. Washington, DC, March, 2002.
   www.sleepfoundation.org/. Accessed: April 15, 2003.
7. Huari, Peter, and Linde, Shirley, *No More Sleepless Nights,* New York: John Wiley and Sons, 1990.

8. National Sleep Foundation, 2003 Sleep in America Poll. [online] available: Washington, DC, 2003.
www.sleepfoundation.org/ (accessed: April 15, 2003).

9. National Sleep Foundation, www.sleepfoundation.org/ (accessed: April 15, 2003).

Development of Competence in Emergency Responders

7

Competent responders are not born into the emergency services; they are developed there. No fire fighter, medic, or law enforcement individual fresh out of training can be expected to have the skills and abilities of a veteran member of the department. Furthermore, everyone increases and develops his or her knowledge and skills at a different speed. In any department, members at every rank have competencies ranging from novice to expert.

When emergency responders change job titles or move to a different station, they may take a step back until their skills develop in response to their new position. In some cases, however, they may find that their competencies do not develop as they would like. This chapter helps emergency responders understand their behavior as they develop competence in the emergency services. Understanding why people act the way they do and how they make decisions can reduce stressful interactions that can occur when emergency responders move along the competency continuum or when many different levels of competency come together in one department.

DREYFUS MODEL OF SKILL ACQUISITION

Competence and skill development have been studied extensively. The Dreyfus Model of Skill Acquisition evolved from a

---

Paul I. Schwartzman, who coauthored this chapter, maintains a private practice as a counselor and consultant in Fairport, New York. Paul became involved with the fire service while developing juvenile fire-setting intervention programs. He also provides consulting services and educational seminars throughout North America on optimizing performance under pressure.

1979 study of the performance of pilots in emergency situations [1] and has since been applied to computers and artificial intelligence [2], development of excellence in nursing practice [3], and other professions. The Dreyfus model has broad applicability to virtually every profession; this chapter applies it to the development of competence in the emergency responder. The model has five stages of development progressing from novice to expert [1, 3]:

1. Novice
2. Advanced beginner
3. Competent responder
4. Proficient responder
5. Expert

As we consider each of the stages of developing competence, keep in mind that all emergency responders—rookies and experienced responders alike—fall somewhere along this continuum. A person's place on this continuum determines how, typically, they make decisions. At times, as discussed later in this chapter, problem-solving methods used at one stage seem at odds with those used at another stage. The conflict that occurs when skills and knowledge do not match the demand of the situation produces stress. Conflicts that arise from different problem-solving strategies are also a source of stress and could be eliminated if the people involved understood how others perceived the issues being confronted.

PERFORMANCE UNDER PRESSURE

The stress response has been well defined in earlier chapters. In this chapter, we'll revisit stress specifically as it relates to performance under pressure. The very nature of emergency services is performance under pressure. Pressure on emergency responders is further magnified when responders apply new skills and knowledge at an emergency scene. This occurs as responders enter each new stage of competency.

Focus of Attention

As discussed in Chapter 1, the stress reaction triggers physical responses that include muscle tension and increased respiratory rate. Psychological or mental changes also occur and one of the most potent of these changes under stress is that attention narrows.

Psychologist Robert Nideffer examined attention under stress as it relates to a person's performance under pressure [4]. His conclusion, simply stated, is that successful performance is the ability to attend to the right thing at the right time. Nideffer's work has been widely recognized and broadly applied in the selection and training of law enforcement and military personnel, including the U.S. Navy Seals [5].

> **Key Point:** *One of the most potent psychological or mental changes under stress is that attention narrows.*

> **Key Point:** *Successful performance is the ability to attend to the right thing at the right time.*

Attentional Styles

In emergencies, the appropriate thing the responder needs to attend to changes, depending on the circumstances and the responder's individual role. To be successful, responders need to shift their attention along two intersecting dimensions, width and direction (see Figure 7.1).

Broad External Attentional Style

A *broad external* focus of concentration is the style used for awareness and sensitivity to surroundings (see Figure 7.2). For example, a fire officer arrives at a fire and conducts a size-up to decide deployment of apparatus, equipment, and fire fighters. A medic uses this focus to perform a "look test" assessment of a patient's overall condition.

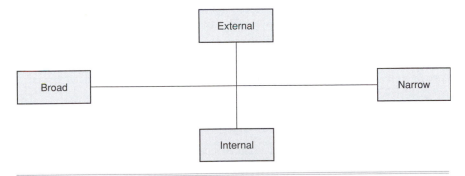

Figure 7.1 *Attentional Styles*

Figure 7.2 *Broad External Attentional Style Needed for Awareness of the "Big Picture" at This Disaster Scene (Source: © 911 Pictures 2003. All rights reserved.)*

Broad Internal Attentional Style

A *broad internal* focus of concentration is used to strategize and to creatively problem solve (see Figure 7.3). An incident commander uses this focus to make adjustments in a rescue plan; a senior officer uses it to develop strategic goals and objectives for his or her company. A president or chairman of the board uses it to develop an annual budget.

Narrow Internal Attentional Style

A *narrow internal* focus of concentration is used to create a logical set of systems and/or procedures (see Figure 7.4). An emergency responder uses this style of concentration to mentally rehearse his or her performance. A manager uses it to create a set of rules or steps that lead to the accomplishment of a goal or objective.

Narrow External Attentional Style

A *narrow external* focus of concentration is used to execute an action to get the job done (see Figure 7.5). This is the type of

Figure 7.3 *Incident Commander Using Broad Internal Attentional Style to Assess Risk at an Incident Site (Source: Photo courtesy of Fairfax County Fire and Rescue Department, Fairfax, Virginia.)*

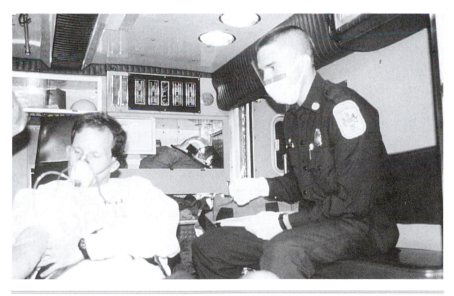

Figure 7.4 *Medic Using Narrow Internal Attentional Style to Review Findings and Mentally Rehearse a Plan of Care. (Source: Photo courtesy of Fairfax County Fire and Rescue Department, Fairfax, Virginia.)*

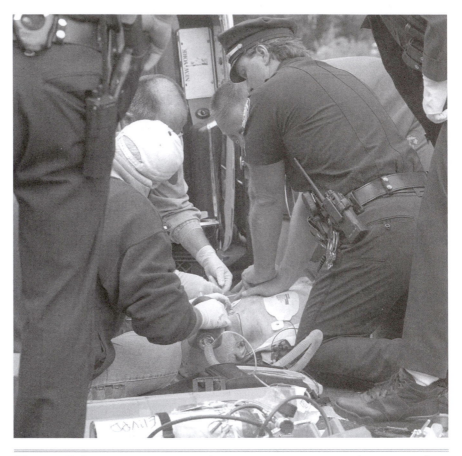

Figure 7.5 *Emergency Responder Using Narrow External Attentional Style to Perform a Life-Saving Action on a Patient (Source: © 911 Pictures 2003. All rights reserved.)*

concentration a medic uses to take a patient's blood pressure or insert an IV. A fire fighter uses this kind of concentration to vent a roof or hit a hydrant. A police officer would use this focus when holding a suspect at gunpoint.

Attentional Style and Level of Competence

You can use only one attentional style at a time. Most people have a dominant attentional style. Under pressure, they will default to their dominant attentional style, even if it is the wrong one for the situation. For example, a rescuer with a narrow external dominant attentional style arriving at a multiple vehicle crash may not shift to the broad external focus needed on arrival, missing important safety hazards such as downed wires and fuel spills. Emergency responders at higher levels of competence, based on their experience, skill, and confidence, are often able to make corrections and shift their focus of attention to perform optimally under pressure.

It is useful to imagine the attentional style as a video camera in your mind's eye. The camera is aimed at what is being attended to. When emergency responders at early levels of competence discover that they are not attending to the right aspect of a situation, their stress response increases, making it more difficult for them to shift to a focus that will be productive. For example, a medic under stress from lack of sleep may have difficulty calculating a drip rate for a drug. The medic's "attentional camera" is likely pointed on a narrow internal focus, emphasizing her physical and emotional sense of exhaustion, and is, therefore, not making the shift to a broad internal focus necessary for performing calculations.

A fire fighter assigned to accountability at a fatal fire scene where a fellow fire fighter has died may be focused internally on his or her own grief reaction rather than externally on the task at hand, and thus have trouble ordering identification tags. A police officer responding to a shooting may experience difficulty finding the location of the call on his roadmap because he is focused internally on rehearsing how he is going to behave at the scene as opposed to externally on reading his map. While his eyes may be directed externally toward the map, his attentional camera is focused internally. Stress may impair the responder's ability to recall dispatch information, vital signs, or details of an incident reported by a witness. These stress-related behavioral impairments may provide an explanation for the seemingly greater numbers of mistakes and injuries that happen with newer, less-experienced emergency responders.

Some individuals who achieve expert status in their initial role may find that they are not as effective in roles with broader responsibility. This sometimes occurs when an expert fire fighter is promoted to an administrative role. He or she may have been extremely proficient as a fire fighter but seems ineffective as a chief. In these individuals, attentional styles may be so ingrained that they are unable to make the shift to a more conceptual style required in an administrative capacity.

This inability to shift attentional styles can also occur within a particular role. Nideffer examined the individual's ability to shift focus of attention in response to changing demands or circumstances, a capacity known as *flexibility*. Some individuals simply are not flexible. It is important for people to be aware of their strengths and weaknesses under pressure. Being less flexible does not mean a responder is incompetent, but it does mean he or she may be better suited for a support role that takes advantage of his or her particular strengths and abilities.

It is also important to note each stage of competence has different priority attentional styles (see Figure 7.6). The priority

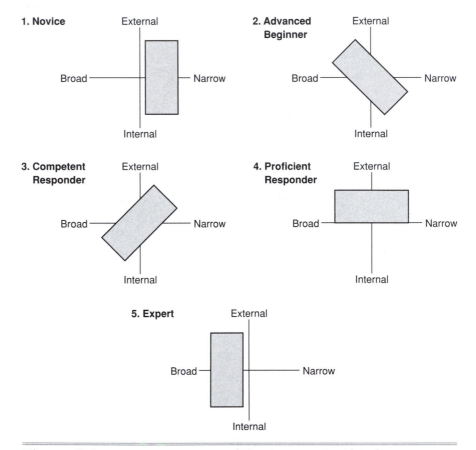

Figure 7.6 *Competency States and Priority Attentional Styles*

attentional styles used at each level of competence reflect the experience and training of that particular stage. They also help to explain why various responders view the same situation in different ways. Decision-making differences between the various levels of competence are affected by attentional styles as well.

STAGES OF COMPETENCE DEVELOPMENT

Stage 1: Novice

The novice is the new kid on the block. Unsure, and with no experience, he or she usually enters a formal department training program or some type of emergency services academy as a trainee. The goal of this initial education is to give trainees an entry ticket that will allow them to participate in actual emergency responses and gain the experience needed to develop their skills. Without experience, the novice is unable to relate most of the information acquired in training to any real events other than simulations and practical exercises in the learning environment.

Role of Rules

The foundation of novice training is a set of rules for novices to follow in each situation they might encounter. This objective- and task-oriented teaching approach produces a novice who thinks in terms of "black and white"; in other words, the novice's learned behaviors are governed by a set of rules and are of a mechanical nature. The novice relies predominantly on a narrow attentional style; his or her success depends on an ability to perform rote mental or mechanical tasks.

> **Key Point:** *The foundation of novice training is a set of rules for novices to follow in each situation they might encounter. They rely predominantly on a narrow attentional style.*

The novice typically prefers to be shown what to do and may express little interest in knowing why things happen. Understanding "why" would shift attention to a broad, internal focus, which at the novice stage is likely to result in a mistake. For example, a novice fire fighter learns to connect a hose coupling using a specific technique. A novice medic learns to start an IV following a set procedure. A rookie police officer is taught to make a traffic stop in a set fashion. Interrupt novices during their performance to query them on how or why they do something, and their routine will often be derailed. Unfamiliarity with their new turf, coupled with efforts to remember all the rules usually make novices slow.

With some training, novices get a foot in the door of emergency services. They acquire a great deal of "book knowledge," but with little or no experience in applying their new knowledge to actual situations, they need to stick rigidly to the rules they were taught. Novices often keep manuals, field guides, books, or computers close at hand to remind them of the rules they need to follow. The rule-governed behavior of novices is limited and inflexible. Rules don't allow for every possible situation that could occur. Novices who follow sets of rules don't use their discretion when making decisions.

Key Point: The rule-governed behavior of novices is limited and inflexible.

Following rules is also a poor way of prioritizing tasks, which leaves the novice at a distinct disadvantage in this activity as well. When situations or circumstances occur that cry out for deviating from strict rules or for changing the priority order of tasks, the novice will not recognize them. The novice requires the help and oversight of experienced emergency responders to perform safely and be even minimally effective on the job. Repetition and practice are necessary and should be encouraged and rewarded at the novice stage.

Key Point: Repetition and practice are necessary and should be encouraged and rewarded at the novice stage.

Rookies and students are not the only novices. Emergency services providers or administrators thrust into a situation where they have limited experience may be reduced to the novice level of performance. This is especially true when they are unfamiliar with the goals and objectives required of them or with the tools they have to work with. New situations come with promotions or transfers to different jobs, for example a seasoned fire fighter is promoted to a leadership position or an experienced cop is promoted to a detective position.

Working in a new or unfamiliar environment also can reduce performance to the novice level as when a veteran medic changes employers and finds herself in completely different ambulances than before or a fire academy instructor finds himself having to speak to a local civic organization for the first time. In each of these situations, new and different roles or environments can lower performance back to a novice level. Although responders' aptitudes and attributes contribute to their level of competence, the demands of the situations in which they find themselves play a far greater role.

Stress and the Novice

The newness of everything around novices makes them particularly prone to symptoms of stress [6, 7]. Training officers, leaders, and more experienced emergency responders should learn to recognize stress-related behaviors of novices because they can easily be mistaken for poor performance. These include The following behaviors:

> ***On the Lookout:*** *Learn to recognize stress-related behaviors of novices so you don't mistake them for poor performance.*

- Decreased reaction time
- Reduced attention span
- Decreased perception of environmental clues
- Decreased efficiency in mental processing
- Impaired short-term memory

Rookies are slow not only because tasks and equipment are new to them, but also because of the increased stress that comes from being the new kid on the block. Stress further restricts the already narrow focus of novices and interferes with their ability to recognize and respond to environmental (external) clues, such as an air bottle alarm, a dummy light on the dashboard, or medical equipment beeps and warning alarms. Experienced responders are not only under less stress than beginners, but also have more broadly focused attentional styles. This helps veteran responders be more attuned to visual and audible clues in their environment and able to more quickly process the meaning of these alarms and signals.

Stage 2: Advanced Beginner

Aspects of the Situation

By acquiring some real-life experience, the advanced beginner performs at a marginally acceptable level. Rigid rules

Key Point: *Aspects of the situation are global characteristics of situations recognizable only with prior experience.*

learned as a novice and applied to many different situations now expose patterns that allow advanced beginners to see (or have someone point out to them) some repeated meaningful elements. These features, which Dreyfus called, "aspects of the situation," differ from the rigid rules of the novice. Aspects of the situation are global characteristics of situations recognizable only with prior experience.

The following are examples of aspects of the situation:

- Evaluating whether a fire requires a 1 ¾- or 2 ½-inch hose
- Distinguishing between different bowel sounds (such as normal, hypoactive, and hyperactive)
- Estimating the speed of a vehicle when deciding whether to activate a radar gun

Use of aspects requires a shift in the priority attentional style of the advanced beginner from the novice's narrow exter-

Key Point: *Use of aspects requires a shift in the priority attentional style of the advanced beginner from the novice's narrow external focus to a broader external focus.*

nal focus to a broader external focus. Whereas external focus shifts from narrow to broad, internal focus remains narrow at the advanced beginner stage as the responder continues to search for logical rules that he or she can consistently apply. Unlike rules, however, aspects cannot be made completely objective. Instructors and mentors can make aspects very explicit, but, unlike rules, no one aspect will apply to all situations.

Developing Guidelines

Key Point: *Training of the advanced beginner moves from following rigid rules to developing guidelines, which are principles that incorporate multiple aspects of situations.*

Training of the advanced beginner, therefore, moves from following rigid rules to developing guidelines. Guidelines are principles that incorporate multiple aspects of situations. These guidelines are similar to rules except that the actions they call for are based on recognizing particular aspects of the situation. Considerable time is spent learning to recognize aspects of actual response situations.

One problem with guidelines is that they tend to treat all aspects equally. For example, ventilation should be done as near to a fire as possible. It should also be done as close to the peak of the roof as practical. But what happens when two aspects such as these conflict? Guidelines leave the advanced beginner still unable to adequately prioritize or sort multiple tasks. Without the ability to recognize what is and isn't important, advanced beginners require the support of other emergency responders who function at least at the competent level, in order to work safely in the field. Although they are more than able to perform simple tasks independently, advanced beginners require supervision for complex duties. The references and guides acquired and kept close at hand during the novice stage are often put to use by the advanced beginner.

Stress and the Advanced Beginner

Stress at the advanced beginner stage can come from multiple sources. The advanced beginner becomes cognizant that the black-and-white rules of the novice may not fit all situations. Their newer "aspects of the situation," learned from experience, provide additional uncertainties because they lack the objectivity of black-and-white rules.

> **Key Point:** *The aspects of the situation, learned from experience, provide additional uncertainties because they lack the objectivity of black-and-white rules.*

Also, advanced beginners experience stress when peers functioning at higher levels of competence allow their frustration with the advanced beginner to show. For example, a medic's field training officer (FTO) becomes annoyed that the medic failed to recognize cool lower extremities as symptomatic of a leaking aortic aneurysm. The patient's presenting problem was chest pain. What contributed to this mistake?

As mentioned earlier, attention narrows under pressure and focus is not always on the most relevant or appropriate cue. In this particular case, the medic was acutely aware of the chest pain and narrowly focused on this presenting problem. Ideally, as illustrated by the FTO, this

> **Key Point:** *Attention narrows under pressure and focus is not always on the most relevant or appropriate cue.*

symptom should have been perceived as being a part of a more serious problem. If the medic had performed optimally under pressure, he would have shifted his focus from a narrow external focus to a broad external focus in order to perceive all "aspects of the situation" and arrive at an accurate diagnosis.

Advanced beginners rely predominantly on a broad external focus of attention, which allows them to observe multiple aspects of a situation. They then shift to a narrow internal focus of attention to apply sets of learned procedures or guidelines to the aspects they have observed. When aspects of the situation do not mesh with their level of competence, the responder's performance suffers.

Centering

How can emergency responders make corrections in real time when they recognize that they are not attending to the most relevant cues? One technique for refocusing attention is called *centering* [8]. The centering technique produces physical and emotional balance by bringing about the relaxing of muscles, the slowing of breathing, and the focusing of concentration on task-relevant cues or tactics. This technique is designed to counter downward performance cycles and to help prepare for optimal performance under pressure. Centering quickly counters physical bracing (i.e., the tendency to tense one's muscles) and negative thoughts that hinder performance. See Table 7.1 for a sample centering routine the advanced beginner medic in the earlier example could have used.

Key Point: Centering produces physical and emotional balance by bringing about the relaxing of muscles, the slowing of breathing, and the focusing of concentration on task-relevant cues or tactics.

In the earlier example, the FTO, like the medic, did not perform effectively under pressure. Understandably, FTOs must assure patient safety and well-being, but their role is to provide teachable moments as well. By expressing annoyance, FTOs may trigger stress in their student, thereby minimizing the opportunity for learning and success. Good coaches are able to shift their own narrow internal focus (or their annoyance) to a broad external focus of attention. This helps them see the student's frustration and helps them as well to guide the student in shifting his or her focus to the bigger picture.

Table 7.1   *Sample Centering Routine*

- Take a slow deep breath from down in your abdomen.
- Sense your center of gravity or balance (usually 1 or 2 inches below and behind your navel) and physically balance yourself. If you are standing, slightly bend your knees.
- Relax your muscles by saying to yourself, as you exhale, "Relax the muscles in my neck, shoulders, arms, and legs."
- Keep your breathing steady and slightly slower than normal.
- Make sure your weight is evenly distributed and that you are balanced and ready to move in any direction.
- Direct your attention to the task at hand.
- Give yourself a technical or tactical self-instruction that is directly related to successful performance by saying to yourself, "Step back for a moment and look at the 'big picture'."

Stage 3: Competent Responder

As the responder gains experience, the ever-increasing numbers of guidelines and rules can become overwhelming. Lacking a means of determining what is most important in any situation, he or she may begin to wonder how anyone ever masters the job skills.

Formulating Plans

With proper direction, an emergency responder working at the same or similar job for between two and three years will typically reach the competent stage. The competent responder can analytically contemplate a problem and has a sense of what is important as well as what can be ignored. The achievement of competence is marked by a reliance on a broad internal attentional style. This style is defined as the ability to formulate plans in a conscious and deliberate fashion.

> **Key Point:** *The achievement of competence is marked by the ability to consciously formulate plans that filter the many rules and guidelines and help the responder to set priorities.*

Plans filter the many rules and guidelines and help the responder to set priorities by determining the elements of a situation that are important and those that can be ignored. These plans help the responder to achieve a level of efficiency and organization. No one can prepare a list for the competent responder of what to do in every possible situation. Plans are formulated by the responder who consciously sees his or her actions analytically and then executes those plans specifically. A newfound sense of responsibility comes with the use of plans.

Prior to the competent stage, if the rules didn't work, the responder could simply rationalize that he or she had been given poor guidance. When things go poorly for competent responders, however, they feel bad because they see it as their mistake. For example, competent fire fighters who lose a structure often blame themselves, even when the odds were clearly against them. Conversely, when things go well, the competent responder feels a sense of exhilaration unknown to the novice or advanced beginner. A competent police officer who nabs a wanted felon during routine patrol is ecstatic, as is the competent medic who successfully resuscitates a small child pulled from the bottom of a swimming pool.

Competent responders feel a sense of mastery and of being able to handle their job and the many variations that come with it. After considerable struggle and effort, the job finally seems to have some semblance of order to it. Despite this feeling of mastery, people under pressure at the competent level still lack speed, efficiency, and true flexibility on the job. For example, a competent driver negotiating a sharp curve may, after considering a number of variables such as road surface condition, speed, and road grade, decide he is going too fast. He then needs to choose from several options such as letting up on the accelerator, removing his foot from the accelerator altogether, or actually braking. Under pressure, the competent responder, who now expects to be able to understand and figure out a solution, may spend precious time using an internal or analytical focus of attention, thinking about what to do. This process slows the competent responder.

Decision-Making Simulations

> **Key Point:** *Training of competent responders shifts from rules and guidelines to decision-making simulations that provide the responder with practice in responding to more complex types of calls having multiple problems.*

Training of competent responders shifts from rules and guidelines to decision-making simulations that provide the responder with practice in responding to more complex types of calls having multiple problems. The competent responder can function independently in most situations and represents the department well in the community. Competent responders are capable of limited leadership roles within the department.

Stress and the Competent Responder

Stress at the competent level often arises from the lack of flexibility seen at this stage. For example, a competent responder faced with an obnoxious citizen at an emergency scene may have difficulty bending his or her activities to accommodate the unique needs of the scene. Further stress accumulates once a competent responder formulates an action plan and is subsequently required to make changes after setting it in play. The centering technique described in Table 7.1 can be helpful to competent responders who find themselves experiencing stress from demands to shift their focus or to change plans at an emergency scene.

Stage 4: Proficient Responder

The proficient emergency responder has learned from experience what to expect in a given situation and how to tailor strategies in response to these happenings. Typically, it takes a responder between three and five years of working in the same job to reach the proficient stage of competence.

Maxims

Operationally, proficient responders see situations as a whole instead of by rules, aspects, guidelines, or plans. They can readily shift their priority attentional style from a broad to a narrow external focus. They not only see the big picture at a scene but are also acutely sensitive to small details that others may miss. What previously was a thoughtful prioritization of situational elements becomes less complicated and more obvious to the proficient responder. Aspects begin to stand out as important, and goals become more obvious.

The proficient responder's performance is guided by *maxims*, which are nearly invisible parts of a situation that have different meanings at different times. Illustrating maxims is challenging because they have meaning only to proficient and expert responders and their meanings vary depending on the circumstances. Maxims might include, for

> **Key Point:** *The proficient responder when guided by maxims—nearly invisible parts of a situation that have different meanings at different times—sees situations as a whole instead of by rules, aspects, guidelines, or plans.*

example, a slightly sagging roof or a minute structural shift that signals potential building collapse at a fire scene. A high heart rate that subtly points to fever and infection could be a medical maxim. A barely noticed behavior that indicates a possible hidden weapon may be a law enforcement maxim. Collectively, maxims vary from situation to situation. Their definitions are elusive, and they primarily serve as a means of highlighting the most important aspects of a specific problem.

From experience proficient responders have developed the ability to survey the whole situation. They rely on a broad external focus of attention, pick out the most important aspects, and recognize when the usual outcomes are not materializing. Before responders achieve the proficient level of competence, they lack the ability to pick out the most important aspects of a situation. The proficient responder, who considers far fewer options than would the advanced beginner or the competent responder, homes in on the most critical elements or aspects of a situation. In fact, actions that previously required deliberate, calculated thinking become intuitive and almost reflexive.

The proficient driver negotiating a sharp curve, for example, realizes intuitively that she is going too fast and immediately either brakes or lets up on the accelerator. By not spending precious time considering all the variables, the proficient driver is far more likely to make the curve safely than is the competent driver who first considers road surface, speed, and road grade before deciding if the vehicle's speed is excessive or not and then decides whether or not to brake. The attentional style of proficient responders shifts from a broad, external focus in which they "sense" the circumstances to a narrow, external focus of attention used to execute a response or direction.

Case Studies

Key Point: *Case studies containing the same level of complexity and ambiguity as real-life scenarios provide the most effective learning situations for proficient responders.*

Proficient responders have very different educational needs and interests. They learn best from case studies that challenge their abilities to handle various situations. The rules and theories used to teach the beginner have little value at the proficient level because they fail to account for the subtleties and nuances (i.e., maxims) that guide the proficient responder's decisions. Attempts to teach proficient responders using rules will probably cause them to counter with situations where the rules would not apply. Case studies containing the same level of complexity and ambiguity as real-life scenarios provide the most effective learning situations for proficient responders.

Early Warning Signal

Benner describes an interesting talent, called the "early warning signal," found most often at the proficient level [3]. It seems that the proficient responder is most likely to recognize a situation going bad before it becomes obvious to others. Proficient fire fighters recognize hazards and problems at a fire scene before they are evident to others. The proficient medic senses a patient's problem before any changes in vital signs occur. Proficient police officers sense something is not quite right during an interview when others are unaware of a problem.

Proficient responders are highly productive, much more so than the average responder. They are known in their departments as excellent problem solvers and receive compliments from coworkers, peers, and customers. Their performance is a benchmark for others. They are often teachers and coaches to other members of the emergency services.

Stress and the Proficient Responder

Stress at the proficient level can result from having the inclination to sound an "early warning signal" but not having sufficient data to convince others to act on what may be considered merely intuition. Although the proficient responder is likely more accurate in decision making than responders at less competent levels, the limited data on which maxims are based hold the potential for error if misinterpreted. With the seriousness of decisions affecting lives and property, mistakes in this arena can be costly as well as stressful.

Stage 5: Expert

Intuitive Responses

The expert responder moves beyond the use of rules, guidelines, or maxims to connect his or her understanding of a situation to the appropriate action. A wide and varied range of experiences

Key Point: *In the expert responder, a wide and varied range of experiences replaces reasoned responses with purely intuitive answers and action.*

replaces reasoned responses with purely intuitive answers and action. The proficient responder sees what needs to be done but must then decide how to accomplish it. The expert responder not only sees what needs to be done but simultaneously sees how to do it. The expert responder immediately sees solutions to problems without having to consider multiple elements of the situation. The priority attentional style associated with expert level responders is a broad external focus coupled with a broad internal focus. Experts base their decisions on selected pieces of the most important information gathered from a broad view of a situation. Less experienced responders use more expansive ranges of information that yield less consistent decisions. Indeed, the performance of expert responders seems so immediate and fluid that it is often extraordinary. Immediate, intuitive responses that distinguish expert from proficient responders come from vast experiences in similar situations, all requiring different tactical actions. These develop in the expert an ability to subclass different situations into categories that all require similar actions.

Experts, however, don't abandon analytical thinking. In fact, experts utilize very advanced analytical skills in situations for which they have no prior experience and in circumstances where they formulate the wrong impression, only to find things not progressing properly as a result. Recovery from incorrect actions taken because of a mistaken impression requires sophisticated use of analytical troubleshooting.

Trying to describe expert performance is a challenge. The deep understanding that experts develop often defies their own ability to explain how and why they make the decisions they do. What must be done is simply done. The sheer number of different situations experts seem capable of recognizing, apparently without any awareness, is mind-boggling. It has been estimated that a master chess player can distinguish between some 50,000 different types of positions [9].

When experts are asked to analyze a hypothetical situation for decision points, they often will say, "Well, it all depends . . . ." More study is needed to elucidate the processes by which expert responders act. Dreyfus summarized expert performance by noting that the expert pilot is no longer flying the plane, but that, instead, he or she is flying [1].

A fire fighter who arrives at the scene of a fully involved structure 2000 feet away from the nearest hydrant and within moments completes a size-up and formulates a complex plan to deliver water using the next arriving apparatus is demonstrating expert performance. A sole medic who on a routine nursing home transfer that suddenly becomes anything but routine delivers two defibrillations, starts an IV, and administers multiple medications to resuscitate her patient, all seemingly without difficulty or distress, is exhibiting expert performance.

Contribution of Experts

One clear contribution of experts to the field of emergency services is their ability to present to others their vision for what is possible in the field [3]. Educational and motivational talks by experts present visions of excellence that challenge and motivate proficient and competent responders to develop their own expertise and progress to higher stages of competence themselves.

Expert responders have an enormous scope and depth of knowledge. Their performance is well beyond most others in the department and they are the leaders of specialized and highly technical teams and task forces. Their leadership skills are exceptional and they often teach and coach others.

Stress may arise when expert responders are asked to mentor or precept rookie members. Because experts tend to think in "gray" terms, they may appear incompetent in the black-and-white world of the rookie. Conversely, unaware that competence develops along a continuum, experts find themselves readily frustrated when they attempt to teach intuitive decision-making skills to members lacking the requisite situational experience.

DEVELOPMENT OF COMPETENCE

Mentoring Emergency Responders

People at different stages of competence exhibit vast differences in their behavior and attentional styles. They rely on different information in the ways they make decisions and they also learn very differently. These differences allow emergency responders to recognize not only where they are currently, in terms of competence and skill and knowledge development, but where others in their department are as well. This insight holds particular significance for mentoring emergency responders through the ranks of the service. Mentors should recognize that rookies usually function at the novice and advanced beginner level.

> **On the Lookout:** *As a mentor, you should recognize that rookies usually function at the novice and advanced beginner levels.*

Situations for novices are black and white, never gray. They are probably not interested in why a certain task needs to be done, only in how to do it. How rookies function contrasts starkly with how experts function. Experts don't use rules but instead act intuitively, based on their experience of thousands of incidents. Consider how the expert and novice face an unfamiliar situation. Whereas the novice thoughtfully and sequentially selects a course of action based on a set of rules and instructions, the expert does none of these. Instead, the expert assesses the overall situation, instantly analyzes the nuances of the situation, and initiates multiple actions that are virtually automatic. Experts, therefore, are probably not good mentors, field training officers, or preceptors. They don't remember the rules they learned in school, and asking them to mentor beginners requires them to regress to the beginner level in order to remember rules they no longer use. Competent responders are probably the best mentors since they are closest to the novice and advanced beginner and utilize the same type of reasoning.

> **Key Point:** *Competent responders are probably the best mentors since they are closest to the novice and advanced beginner and utilize the same type of reasoning.*

Multiple Training Approaches

Moving through the stages of competence requires different learning environments. This is important for emergency service leaders to know, for they sometimes believe that a single educational approach is adequate for all levels of responders.

Traditional Curricula

Traditional schools, colleges, and academies bring students from the novice through advanced beginner and competent levels with curricula focused on facts, rules, principles, procedures, and psychomotor skills.

Experiential Learning

Raising competence to higher levels requires a far more intensive approach than is used in the traditional educational environment. Experiential learning using apprenticeships and personal coaching are necessary for responders to advance to the proficient and expert levels of competence. It is fallacy to believe that progression to higher levels of competence comes only to those with greater book knowledge. In truth, graduates are more aptly judged on performance than on book knowledge. Experts and masters achieve their standing by learning from their years of experience, continual practice, and frequent performance under the watchful eyes of others.

Steps to Advancement

Advancement to the expert level of competence and beyond requires a lot of self-direction and some very good teachers. Suggested steps to advancement include the following:

- Seek out experts in your field and join or work with their departments so that you can learn how they think and act.
- Remember that continual practice is critical to growth and development.
- Because professional affiliations are important, network with others, attend conferences, and listen to the strongest and most respected speakers.
- Seek out the opinions of others on your performance and learn to incorporate their feedback into personal and professional improvements.
- Look for excellent practices and incorporate them into your own style.
- Keep in touch with the experts you most admire for enduring fountains of inspiration.

SUMMARY

Not everyone is destined to reach the expert stage of competence as an emergency responder, nor does every responder even care to reach beyond the competent stage. Recognizing at which stage of competence you, your coworkers, and your leaders are will provide keen insight into behavior, decision making, and learning needs. Responders at each stage of competence see their world and each problem they encounter in very different ways, based on their different priority attentional styles. Understanding these differences can make working together easier and far more comfortable than blithely assuming that every responder will think and act alike.

REFERENCES

1. Dreyfus, Stuart E., and Dreyfus, Hubert L., *The Scope, Limits, and Training Implications of Three Models of Aircraft Pilot Emergency Response Behavior.* Unpublished report supported by the United States Air Force Office of Scientific Research (Grant AFOSR-78-3594), University of California at Berkeley; February 1979.
2. Dreyfus, Hubert L., *What Computers Can't Do: The Limits of Artificial Intelligence,* rev. ed., New York: Harper & Row, 1979.
3. Benner, Patricia, *From Novice to Expert: Excellence and Power in Clinical Nursing Practice,* Commemorative ed., Upper Saddle River, NJ: Prentice Hall, 2001.
4. Nideffer, Robert M., *The Attentional and Interpersonal Style Inventory: Theory and Applications,* New Berlin, WI: Assessment Systems International, 1992.
5. Nideffer, Robert M., *Predicting Human Behavior,* New Berlin, WI: ASI Publications, 1999.
6. Asken, Michael J., *Psycheresponse: Psychological Skills for Optimal Performance by Emergency Responders.* Upper Saddle River, NJ: Prentice Hall, 1993.
7. Lieberman, H. R., et al., *The "Fog of War": Documenting Cognitive Decrements Associated with the Stress of Combat.* U.S. Army Research Institute for Environmental Medicine. Oral paper presented at the 23rd Army Science Conference, Orlando, Florida: December, 2002.
   [on-line] available: http://mit.ucf.edu/muri/conf/ 23rdarmyscience/manuscripts/I/IO-01.PDF (accessed April 30, 2003).
8. Nideffer, R.M., Psyched to Win, Champaign, IL: Leisure Press, 1992.
9. Dreyfus, Hubert L., and Dreyfus, Stuart E., *From Socrates to Expert Systems: The Limits and Dangers of Calculative Rationality.* University of California at Berkeley: Winter 2002. [on-line] available: www.psych.utoronto.ca/~reingold/courses/ai/cache/Socrates.html (accessed May 4, 2003).

Personal Stress Management Program

8

E mergency services demand a responder's time, energy, and attention. The job of the emergency responder contrasts starkly with most other jobs, which explains why successful responders seem to have been cast from a different mold than ordinary citizens. Throughout this book we've looked at the effects of stress—both positive and negative—as well as the unique ways stress impacts emergency responders. Coping strategies for many of the specific stressors emergency responders encounter were included to help responders recognize the pitfalls and steer away from them when they encountered them. In the end, however, no book can give you a strategy to deal with every stressor you might encounter. Stress in your life, either from an isolated event or from many things combined, will challenge you to cope. Not all of these stressors have been specifically mentioned nor could every possible source of stress be addressed. Coping with stress requires each of us to have a personal stress management program. This chapter looks at the ingredients common to many programs and provides emergency responders with a guide to developing their own personal stress management program.

OVERVIEW OF STRESS MANAGEMENT

Types of Stress Management Programs

There are innumerable personal stress management books by hundreds of authors, each of whom has a theory on stress management that offers you a happy and healthy life in a stressful world. Comparing stress management models can be overwhelming. They range from art therapy to biofeedback, diet and nutrition to exercise and fitness, meditation to relaxation techniques, and time management to yoga. Yet, as mentioned in Chapter 3, stress management models are like pants: No one pair fits everyone. Think about this, for example, in the simple terms of a vacation. Some people would consider mountain climbing in the Swiss Alps an extremely relaxing vacation and a wonderful stress management tool. Others would experience stress just thinking about such a trip, preferring instead to take a week off from work to put a new roof on their house. Still others consider lounging around the pool and reading a book as the perfect vacation. Unfortunately, stress management models are also like religion. When friends or coworkers find a model that produces genuinely beneficial results for them, they may try to convince others to use the same method, believing that others will experience the same results.

> *Key Point:* Because we are all different, each of us needs a personalized program for stress management that takes into account the many facets of our individual and unique body, mind, and life situation.

A key point emphasized throughout has been that different people are affected by stress in different ways. Because we are all different, each of us needs a personalized program for stress management that takes into account the many facets of our individual and unique body, mind, and life situation. Another key point has been that emergency responders share certain personality characteristics, which, when combined with their work, cause them to experience certain common stressors.

When developing a personal stress management program, emergency responders can choose from a wealth of models. Keep in mind, however, that rarely will one single model suit all of an individual responder's needs. The average emergency responder will most likely need to choose parts of various models and blend them together into his or her own personal stress management program.

> *On the Lookout:* You will most likely need to choose parts of various models and blend them together into your own personal stress management program.

Maslow's Hierarchy of Needs

Psychologist Abraham Maslow observed from his work with monkeys that certain needs seemed to take precedence over others. In 1943, he developed a paradigm called the *hierarchy of needs*, in which he rank-ordered human needs from the most important physical needs to secondary psychological needs (see Figure 8.1). This 60-year-old hierarchy is still considered relevant today. Survival is

Key Point: *Maslow's hierarchy of needs rank-orders human needs from the most important physical needs to secondary psychological needs.*

predicated on meeting certain basic physical needs as well as those of safety, order, and security (see Table 8.1). Basic physical needs take priority over all others and include air to breathe, food, fluids, and sleep. Safety needs are secondary only to physical needs and include clothing, housing, money, employment, and a stable support system of others whom we trust and feel safe with [1].

Coping with stress can only come after these basic needs for survival are fulfilled. Just as you wouldn't discuss the latest movie you saw with a patient suffering from a complete airway obstruction, or converse with a homeowner about your last vacation while his house is fully engulfed in flames, so must you put first things first as you cope with stress. This same principle forms the basis for the "psychological first aid" now recom-

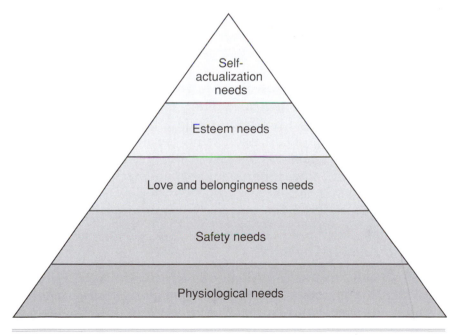

Figure 8.1 *Maslow's Hierarchy of Needs*

Table 8.1   *Putting First Things First: Survival*

Physical Needs	Safety Needs
Air	Clothing
Fluids	Housing
Food	Money
Sleep	Employment
	Support of others
	Feeling safe

mended for people exposed to extremely stressful events. Coping with stress can only come after these basic needs for survival are fulfilled.

Let's start by considering the three realms of a person's life experience where demands—and one's perceptions of those demands—can cause stress: body, mind, and interactions with others. As we consider each, we'll also look at how they might differ for the emergency responder. Although the body, the mind, and our interactions with the world around us aren't completely independent of each other, considering them each as a separate realm is a useful way to categorize the many tools at our disposal for stress management.

THE BODY AND STRESS

Our body is intricately involved in the stress response, both chemically and physiologically. The general adaptation syndrome (GAS) described in Chapter 1 details the response of our body to stress. Our body itself can also become a source of stress, especially when it is not physically able to respond to the demands placed on us. Emergency responders have some very exceptional demands that are seen in few other occupations or walks of life.

Physical Activity Demands

The first and most serious demands are the extreme shifts in activity often placed on emergency responders. There are other occupations that require intense physical activity, but none that requires it instantaneously without warning. Yet that is exactly how fire fighters, medics, and police officers respond. One minute they are resting (at times even sleeping) and the next moment they are asked to shift to a total high-intensity physical and psychological state. That this physical activity demand is the recipe for injuries and failure of body parts is readily apparent. Even a highly conditioned athlete would not consider accelerating to peak performance without a warm-up period; yet emergency responders routinely move from zero to warp speed in the blink of an eye [2].

Good Body Mechanics and Physical Conditioning

The potential for an emergency responder's body to fail under routine job stress is tremendous. Thus a wise stress prevention tool for emergency responders is to maximize strength, flexibility, and endurance and to develop an excellent awareness of good body mechanics. Injury prevention through the use of back supports and proper body mechanics when lifting and moving heavy objects or people is stress prevention.

> **On the Lookout:** *Maximize your strength, flexibility, and endurance and develop your awareness of good body mechanics.*

Unfortunately, emergency scenes are not controlled environments. Medics will always have to lift heavy people in narrow hallways; police officers will always need to run after and capture criminals; and fire fighters will always need to manipulate rescue tools in tight, difficult-to-maneuver spaces. In many of these situations, the physical conditioning of the responder can be just as crucial in preventing injuries as are good body mechanics.

Benefits of Regular Exercise

Secondary benefits of physical conditioning extend to anyone who exercises regularly and these benefits are wide-ranging (see Figure 8.2). Studies of the benefits of regular exercise have proven it a sensible investment in a person's health and well-being (see Table 8.2) [3]. The definition of regular exercise has changed somewhat over the years as more studies have been conducted on the amount of exercise needed for benefits to accrue. Currently, expert consensus defines regular exercise as at least 30 minutes of moderate level activity for most days of the week [4]. With a regular program of exercise, benefits become

Figure 8.2 *Weight Lifter Exercising (Source: © Rachel Epstein/The Image Works)*

evident after between six and eight weeks of training in most people. Of course, every individual is different, and benefits gained will vary with the type, difficulty, frequency, and duration of exercise. Emergency responders are likely to have different needs for conditioning based on their specific job duties. A police detective would probably choose a somewhat different conditioning program than would a fire fighter medic.

Noise

Another stressor that especially targets emergency responders is noise. Virtually no research has looked specifically at the effects of noise on stress levels of emergency responders, but it would be fair to say that the emergency services environment can be very loud at times. Studies have shown that repeated exposure to high-decibel noise not only activates a stress response in the body but also produces hearing loss [1]. Consider the many

> **Key Point:** *Studies have shown that repeated exposure to high-decibel noise not only activates a stress response in the body but also produces hearing loss.*

Table 18.2   *Benefits of Regular Exercise*

- Decreased resting heart rate
- Lowered blood pressure
- More efficient cardiovascular system
- Less muscle tension
- Increased strength
- Improved endurance
- Better quality of sleep
- Improved resistance to colds and illness
- Improved blood cholesterol levels
- Decreased body fat and better appetite control
- Stronger bones
- Slower aging
- Improved psychological outlook
- Reduced anxiety
- Greater feelings of self-esteem
- Decreased sensitivity to temperature extremes, noise, and pain
- Better performance under stress
- Less nervous response to stress

Source: Adapted from Michael J. Asken, *Psycheresponse: Psychological Skills for Optimal Performance by Emergency Responders*. Upper Saddle River, NJ: Prentice Hall, 1993; and National Institutes of Health, *Energize Yourself! Stay Physically Active*. National Heart, Lung, and Blood Institute and Office of Research on Minority Health.

sources of noise in the environment of the emergency responder. Also consider the fact that some emergency responders feel physically and mentally "ready to go" during the entire time they are on duty and that the many noises probably add to already elevated levels of excitement. To reduce exposure to noise, emergency responders could use hearing protection when appropriate, perhaps even while sleeping to reduce the "shock effect" of station alerting systems.

> ***On the Lookout:*** *Use hearing protection when appropriate to reduce exposure to loud noises, perhaps even while sleeping to reduce the "shock effect" of station alerting systems.*

Food

Food is often overlooked both as a stressor and a stress management tool for emergency responders. Nutrition has a direct relationship to stress in that the more stress your body is under, the higher your requirements for good nutrition are. Nutrition is also a source of stress to people. Emergency responders could become stressed by poor nutrition for several reasons. Shift work often limits available food choices and prevents emergency responders from eating at home. Some emergency responders spend their entire time on patrol or stationed in the field where cooking facilities are not available.

> *Key Point: Emergency responders can improve their health, feel better, and control their weight more effectively when their diet is nutritionally sound.*

In addition meal breaks may be few and far between. Emergency responders can improve their health, feel better, and control their weight more effectively when their diet is nutritionally sound.

In addition to being without adequate cooking facilities or too busy to use them, emergency responders also find themselves needing to vary their nutritional intake based on intense levels of physical activity coupled with rest. Healthy foods that can be carried and stored in virtually any environment include bread, crackers, granola bars, cheeses, peanut butter, nuts, dried fruits, beef jerky, and salami [5].

For periods of exceptional exertion, energy bars provide a boost of carbohydrates, usually accompanied by some fat and protein as well. Compared to other sources of energy, carbohydrates are the easiest for the body to convert to energy so they make a particularly effective pick-me-up for emergency responders as well as for athletes and mountain climbers [5]. Vitamin supplements may also be of value to emergency responders, but are beyond the limited scope of this book.

Fluids

Hydration is also important for responders to keep in mind, especially when their work causes them to perspire a great deal. During strenuous activity, losing more than 5 quarts of water daily through sweat and evaporation of moisture from breathing is not uncommon [5]. Drastic effects on performance and well-being

> *On the Lookout: Hydration is also important for responders to keep in mind, especially when their work causes them to perspire a great deal.*

can result when losses are not replenished. Emergency responders could reduce stress by developing a sense of their own needs for replenishment of fluids and nutrition in order to maintain an optimal level of performance at all times.

Sleep

Sleep, already covered extensively in Chapter 6, deserves to be mentioned here. A comprehensive stress management program for emergency responders recognizes the importance of sufficient good-quality sleep in preventing stress and maintaining overall health.

> **On the Lookout:** *A comprehensive stress management program for emergency responders recognizes the importance of sufficient good-quality sleep in preventing stress and maintaining overall health.*

THE MIND AND STRESS

Power of the Mind

The mind is the next realm frequently in need of stress management strategies. The mind forms reactions to the various situations, people, and things around us, transforming some of these from mere events to stressors. Because the mind is a very powerful place, many of the mainstream stress management models focus on that realm. Norman Cousins eloquently depicted the power of the mind in his book *Anatomy of an Illness*, in which he describes harnessing the amazing power of his own mind to fight, and triumph over, a debilitating disease [6]. You can learn to harness the power of your mind as well by deciding how you will react to the events and stressors you meet in life. One of the most powerful messages of this book is that you can let yourself be controlled by people, things, and events in your life—or you can take charge of your reactions to the world around you.

Rest

Rest is a destressing tool that many stress management models espouse. The ability to rest can be a difficult skill for emergency responders to incorporate into their lives. The reasons probably result from two personality traits common to responders: (1) their extreme dedication to their work, which leads them to work more than the usual number of hours in a typical work-week and feel uncomfortable when away from their home territory, and (2) their action orientation, which makes the very idea of rest and relaxation sound extremely boring. In spite of this, even emergency responders need a break from their work and their busy lives in order to refresh and regenerate their minds.

Diversions and Distractions

Diversions and distractions are forms of stress management for the mind.

Hobbies

A hobby is a form of stress management that allows some variety and difference in life. Hobbies are distractions completely under the control of the individual. They allow emergency responders to change their scenery, to promote clearer thinking, to increase self-esteem, and to develop character and confidence [1]. Contrary to popular belief, however, not all hobbies meet the conventional definition of "complete relaxation." Emergency responders can find hobbies that meet their needs for action and adventure, such as parachuting, skydiving, hang gliding, rock or ice climbing, and engaging in competitive sports [1] (see Figure 8.3). Owning a pet can also be considered having a hobby.

Humor

Humor is a form of distraction that can be used to relieve stress. Emergency responders are probably most familiar with "black humor" or "gallows humor," which allows them to become more comfortable as they deal with brushes with death. Norman Cousins believed that good emotions resulting from humor had positive effects on the body and had the ability to heal [6]. Laughter is a way for people to balance emotions in their lives [1].

Journal Writing

Journal writing is a technique for dealing with stress that may have particular usefulness to emergency responders in certain

Figure 8.3 *Rock Climbing (Source: © Jim West/The Image Works)*

situations. Workshops on journal writing teach people to express their thoughts and emotions on paper. The result is an improved ability to express oneself and an enhanced awareness of thoughts, feelings, and insights [1]. Emergency responders experience difficulty at times trying to get particularly stressful scenes out of their mind, as we discussed in detail in Chapter 4. Journal writing is a technique that can help people transfer difficult events

> *Key Point: Journal writing is a technique for transferring difficult events from the mind to a piece of paper.*

from the mind to a piece of paper. Committing difficult events to paper eases the need to keep them forever in the mind and can reduce some of the lingering stress associated with catastrophic happenings.

Spirituality

The role of spirituality cannot be ignored in stress management. It is common for people to ponder over the meaning of life and their reason for living. Religion or spirituality offers a connectedness to the earth, to others, and often to a higher power as well. The soul, which represents a person's spiritual existence, has a direct relationship to the mind and body. Spirituality provides many people with the tools to effectively deal with many of life's stressors [1].

Relaxation Techniques

Many stress management models incorporate relaxation techniques into their programs. Emergency responders can benefit greatly from relaxation techniques to gain control over the adrenaline rushes they experience. Regular use of relaxation techniques allows emergency responders to improve their performance at emergency scenes by improving their ability to focus attention and energy. Responders who regularly use relaxation techniques also have increased resistance to stress [2]. In view of the fact that there are so many sources of information on relaxation techniques available to emergency responders who might consider using them, we'll limit our discussion here to a brief review of the most common [1].

Breathing Techniques

Breathing techniques include, for example, diaphragmatic breathing and deep breathing.

Meditation

Types of meditation include, for example, transcendental meditation (TM), Zen (Zazen) meditation, mindfulness techniques, the relaxation response.

Yoga

Over a hundred different schools or types of yoga are taught today. All involve combinations of exercises (especially stretching exercises), breathing techniques, and meditation.

Mental Imagery / Visualization

Those who practice this relaxation technique use the imagination to visualize peaceful scenes from nature, or to see themselves acting in a new or different fashion when facing a stressor.

Music Therapy / Art Therapy

Those who practice these relaxation techniques use music or art as a relaxation technique.

Massage Therapy

Massage therapy uses human touch to produce muscle relaxation and improve general well-being. Types of massage therapy include, for example, shiatsu, Swedish, rolfing, myofascial release therapy, sports massage, Traeger, zero balancing, postural integration, craniosacral therapy, Reiki, on-site massage, reflexology, and trigger-point therapy.

Martial Arts

The martial arts work to integrate the body and mind with the universe. They include, for example, tai chi chuan, tae kwon do, aikido, tae bo, karate, jujitsu, ba-qua, and hsing-i.

Progressive Muscular Relaxation (PMR)

PMR is a technique developed in 1939 by Edmund Jacobson to reduce muscle tension (the most common physical symptom of stress), thereby preventing cumulative effects of stress on the body. PMR has shown particularly positive benefits when used by emergency responders. It also offers the advantage of achieving relaxation in very short periods of time, once the technique is learned [2].

A quiet environment is needed to initially learn and practice PMR. Start in a comfortable position, such as sitting in a recliner. Begin by taking several slow, deep breaths, holding for several seconds before exhaling. PMR entails alternately tensing then relaxing the various muscle groups of the body. You might begin by closing your eyes very tightly, holding for 2 or 3 seconds, and then relaxing. Breathe in while tensing the muscles and exhale while relaxing. Next you might tense and relax the muscles of your jaw, followed by the neck, shoulders, biceps, and so on.

PMR also involves enhancing relaxation by choosing and repeating a special cue word during each relaxation cycle. Emergency responders gain the best advantage from PMR by choosing cue words relevant to performance under pressure such as *focus*, *relax*, and *easy* [2]. To learn PMR, you will need to write down or record the routine you develop and practice it repeatedly. The basic process takes about 15 minutes. Once learned, a shorter form can be employed, using only your cue word and the relaxation phase (muscles are no longer tensed first). PMR relaxes the muscles and, in turn reduces your stress response. It will also increase your awareness of muscle tension, and help you to condition your mind and body to relax.

Centering

Based on the discipline of the martial art aikido, centering is a breathing and refocusing technique used to relax and control performance under pressure. Developed by psychologist Robert Nideffer, centering was previously described in detail as a technique useful to emergency responders.

Autogenic Training

This technique teaches self-regulation of specific body functions such as circulation, blood pressure, and heart rate.

Biofeedback

Biofeedback is a relaxation technique that utilizes monitoring instruments to provide feedback to users about their physical stress responses. With feedback, users learn to control their stress responses. Due to the direct feedback mechanism, biofeedback is considered one of the most effective stress management tools [1]. An example of a relatively simple form of biofeedback would be a heart rate monitor worn by a runner or a person exercising (see Figure 8.4).

Hypnosis/Self-Hypnosis

This technique places a person in a highly suggestible state. Popular uses include smoking cessation, weight loss, reducing phobias, and anxiety problems. Hypnosis also has been used with success to improve test-taking ability, sports performance, sleep problems, and to decrease the amount of pain medication needed during and after surgery.

Figure 8.4 *Person Checking Wrist Heart Rate Monitor While Exercising (Source: Courtesy of Polar Electro Inc.)*

INTERACTION WITH OTHERS AND STRESS

Our interactions with others form the third and final realm of our lives that can lead to stress. As we have seen in statistics presented previously, the toll of stress-related problems seem to disproportionately affect emergency responders when it comes to interactions with others. There are resources available for people who find stress affecting their ability to have happy and healthy connections with family and friends. This assistance usually comes under the subjects of time management and scheduling, money management and financial planning, employment or career counseling, individual or family counseling, or a wide variety of communication skills for use in being more assertive, dealing with difficult people, conflict management, problem solving, listening, and values clarification, to name a few.

Inner Voice

Before considering some of the specifics of stress management pertaining to our interactions with others, lets consider one additional prerequisite beyond Maslow's hierarchy of needs mentioned earlier. That additional prerequisite is to listen to the voice inside yourself. In his book on fear, noted violence consultant Gavin de Becker uses a series of interviews to dramatically illustrate how people tend to ignore predictions and warnings that arise from within themselves. He encourages readers to nurture and listen to their intuition and instincts rather than to suppress and deny them.

Like the many victims of violent crime interviewed by de Becker [7], we often ignore our inner voice, telling us something is not right here. Think about some of the things you do regularly, like that cup of coffee you perched on the roof of your car, ignoring your inner voice warning you, "it's going to spill." And knowing, after it did spill, that you should have listened. Some mistakes are small, others much larger, such as staying in a job that your heart knows is a dead end; spending increasing amounts of time away from home, ignoring the signals of a crumbling relationship; or getting drunk repeatedly with your eyes closed to ever-increasing isolation. These are actions that will inevitably lead to greater stress and larger consequences. Emergency responders should learn to listen to their inner voice.

> **On the Lookout:** *You must learn to listen to your inner voice.*

Time Management

In earlier chapters we saw how emergency responders may develop a tendency to spend less and less time at home. Time, or the lack of it, is a problem not only for emergency responders; it is often rated by the general public as being a leading stressor in their lives [1]. Creative strategies can help us manage time, but a day will never be longer than 24 hours and a week never greater than 7 days. Better time management seminars, books, and systems all teach a coping strategy called *social engineering*, in which individuals learn to reorganize their goals and schedule so that the end result produces less stress.

Unlike other stress management tools, time management recognizes that some stress related to time is unavoidable. The key to good time management is the ability to schedule, prioritize, and meet responsibilities to yourself and others in a manner that is the least stressful. Although many things interfere with effective time management, solutions are often simple. For example, an emergency responder who enrolls in a martial arts class with his daughter is able to spend time with his family while simultaneously improving his fitness. Others solutions may be more complex, involving intricate juggling of schedules and reordering of life priorities.

Financial Issues

Aside from time, people cite money as a limiting factor in recreation and leisure time habits [1]. Money issues and financial security play such a major role in life stress that they rate high on the Holmes and Rahe stress survey included in Chapter 1. Author Gail Sheehy, who wrote several books about predictable life crises in the adult lifespan, often cited financial issues and their effects as being a key stressor [8]. Emergency responders fall on the moderate to low end of the pay curve and volunteer fire fighters may receive no pay at all.

Financial planning and money management are areas in which emergency responders could proactively address what is undoubtedly a significant cause of stress. Courses, seminars, and books are available on this topic, as are financial advisers and planners. To get started in this endeavor, find peers who can advise you on financial planners or money management courses and books they have used.

> *On the Lookout: Find peers who can advise you on financial planners or money management courses and books they have used.*

Counseling

Seeking counseling has often been considered a sign of weakness by emergency responders. Police studies, for example, have demonstrated that even when counseling is offered to officers, many are unwilling to make use of it for fear of ridicule by their peers or because they fear it will do damage to their careers [9]. The purpose of counseling is to help persons improve their success and happiness with some specific aspect of their life. Types of counseling range from employment and career counseling to individual and family counseling. When a person is unable to resolve a stressful situation involving work, friends, or family, seeking help from a professional counselor seems a reasonable action to take. Counselors are to stress-related problems what coaches are to sports.

Many types of counseling are commonly available to emergency responders. Chapter 4 mentioned critical incident stress debriefing (CISD), which is not a form of counseling but should be able to refer responders who desire or need counseling to an appropriate mental health professional. An often overlooked resource is an employee assistance program (EAP). Recognizing that personal problems affect work performance, many employers offer various EAP services to their workers, often at little or no cost. EAP services typically include marriage, family, and individual counseling; conflict resolution and mediation services; substance abuse treatment and counseling; legal and financial counseling; management assistance and training; as well as a variety of services to help employees deal with critical incidents.

Communication Skills

Few people can claim a mastery of communication skills. From this perceived need have sprung seminars on topics ranging listening skills, dealing with difficult people, and problem solving to assertiveness training, communicating with children and teens, and values clarification. Consider an occasional brush up on your communications skills as an element to include in your personal stress management program.

On the Lookout: Consider an occasional brush up on your communications skills as an element to include in your personal stress management program.

Like many other stress management strategies, communication skills are often techniques you have learned and used in the past that may not have stayed with you.

An important element of communication skills as it relates to emergency responders is the concept of encouragement. Encouragement can counter the negativity that can slip into the work and home life of emergency responders. Providing positive encouragement can have immense antistress value both at home and at work. Make an effort to regularly pat your fellow emergency responders on the back, recognizing their unique positive contributions to the department. Congratulate members on the positive aspects of their performance at emergency responses and drills.

> *On the Lookout:* Pat your fellow emergency responders on the back, recognizing their unique positive contributions to the department.

Focus on your own positive aspects, and apply this same degree of positivity to your home life. A positive outlook has tremendous value as a stress reduction tool. Psychiatrist Victor E. Frankl, for example, attributed a positive outlook to keeping him and many others alive during three years in Auschwitz and other Nazi prisons [10].

> *Key Point:* A positive outlook has tremendous value as a stress reduction tool.

Remember that stress management is a dynamic process. Every day brings new stress into a person's life. Emergency responders will need to continually modify their personal stress management program to suit their changing needs. Don't be afraid to try different approaches, you have nothing to lose and much happiness to gain.

SUMMARY

Coping with stress requires emergency responders to have a personal stress management program, much like other people in society do. Particular facets of emergency response work and the personalities of responders produce greater stress in some areas and make some stress management models more useful than others. Current stress management models focus on the three realms in which individuals regularly confront stressors: the body, the mind, and interactions with others. Specific ways in which the emergency responders might differ in using stress management models in each of these realms were considered.

Emergency responders should consider the following five key recommendations when developing their personal stress management program:

1. *Condition your body and your mind.* Because your body must instantaneously move from rest to peak performance, optimal physical and mental conditioning will help you to consistently meet the performance demands placed upon you.

2. *Listen to your inner voice.* Learning to listen to your inner voice can have tremendous benefits as a stress management tool. You, like most emergency responders are probably an astute observer. Your intuition is likely more often correct than you give yourself credit for. Give your intuition the credibility it deserves.

3. *Never stop growing and learning.* The world is ever changing and the stress it provides changes as well. You see good and bad examples of stress management in the people you serve every day. Take advantage by learning from your experiences and seek out new and better ways of doing things.

4. *Be positive.* Those who serve as continual sources of positive encouragement are not only happier themselves, but make those around them happier as well.

5. *Have a life outside emergency services.* Nothing is more isolating and restrictive than the absence of hobbies, friends, and outside interests. Without an opportunity to get away from the emergency services, you lose the ability to recharge your batteries and return to work with fresh perspectives and new ideas.

REFERENCES

1. Seaward, Brian Luke, *Managing Stress: Principles and Strategies for Health and Wellbeing,* 3rd ed., Sudbury, MA: Jones and Bartlett Publishers, 2002.
2. Asken, Michael J., *Psycheresponse: Psychological Skills for Optimal Performance by Emergency Responders*, Upper Saddle River, NJ: Prentice Hall, 1993.

3. National Institutes of Health, *Energize Yourself! Stay Physically* Active, National Heart, Lung, and Blood Institute and Office of Research on Minority Health. N/H Publication No. 97–4059. [on-line] available: www.nhlbi.nih.gov/health/public/heart/other/chdblack/energize.pdf (accessed September 1997).

4. American College of Sports Medicine, *How Much Exercise Is Enough?* ACSM Fit Society Page. Winter 2003. [on-line] available: www.acsm.org/health+fitness/pdf/fitsociety/fitsc103.pdf

5. Graydon, Don, *Mountaineering: The Freedom of the Hills,* 6th ed., Seattle, WA: The Mountaineers, 1997.

6. Cousins, Norman, *Anatomy of an Illness*, New York: W. W. Norton & Company, 1979.

7. de Becker, Gavin, *The Gift of Fear,* New York: Little, Brown and Company, 1997.

8. Sheehy, Gail, *New Passages: Mapping Your Life Across Time,* New York: Ballantine Books, 1996.

9. Morris, Larry A.; Morgan, Michael; and Gilmartin, Kevin M., *National Institute of Justice Law Enforcement Peer Support Training*, U.S. Department of Justice Document # 189124. [on-line] available: www.ncjrs.org/pdffiles1/nij/grants/ 189124.pdf (accessed July 17, 2001).

10. Frankl, Viktor E., *Man's Search for Meaning*, New York: Pocket Books, 1963.

Index

Accidents, 67, 83
Action orientation, 25–26
  impact on home life, 71
  stress levels and, 130
Acute stress disorder (ASD)
  categories of symptoms, 49,
    51–52
  critical incident stress debriefing
    (CISD) and, 53–61, 138
  nature of, 48–49
  responses to symptoms of, 52–53
Adrenaline, role of, 19–21, 25, 32,
  64, 71
Advanced beginner stage of competence development, 98,
  107–111
  aspects of situation, 107–108,
  109
  centering routine and, 110–111
  developing guidelines, 108–109
  stress and, 109–110
Aggression, in conflict management, 38–39
Airplane pilots, 98
Alarm phase, 2–4
Alcohol
  impairment due to, 83, 87
  sleep and, 93
Alcoholism, 53
American Psychiatric Association
  (APA), 48, 49
Analytical troubleshooting,
  116–117
Anatomy of an Illness (Cousins),
  129
Anxiety, 49, 52
Art therapy, 133

Asken, Michael J., 127
Asthma, 6
Attentional styles, 98–105
  broad external attentional style,
    99, 100, 107–111, 113–115,
    116
  broad internal attentional style,
    100, 101, 111–113, 116
  inability to shift, 103–105, 113
  level of competence and, 102–105
  narrow external attentional
    style, 100–102, 105–107,
    113–115
  narrow internal attentional
    style, 100, 101
Attention to detail, 8, 22
Autogenic training, 134
Awareness, in home life, 75

Benner, Patricia, 115
Biofeedback, 134
Blaming, 42
Blood alcohol content (BAC), 83, 87
Body mechanics, 125
Breaking point, 10
Breathing techniques, 132
Broad external attentional style,
  99, 100, 107–111, 113–115,
  116
Broad internal attentional style,
  100, 101, 111–113, 116
Buscaglia, Leo, 34

Caffeine, 92
Cannon, Walter, 4
Case studies, 114–115
Centering, 110, 134

Choice, reaction to stress and,
    14–15, 50, 129
Chronic stress, 5–6
Circadian rhythms
    diet and, 92, 93
    light and, 85–86, 91
    nature of, 84–86
    non-rapid eye movement
        (NREM) and, 86
    overriding, 87–88
    rapid eye movement (REM) and,
        86
Communication skills, 69–77
    for emergency responder, 71–75
    listening skills in, 73–74, 77
    for significant others, 75–77
    in stress management, 138–139
    value of feedback and, 69–70
Competence development, 97–120
    attentional styles and, 98–105
    mentoring and, 117–118
    stages in Dreyfus Model of Skill
        Acquisition, 97–98, 105–117
    training approaches and, 118–119
Competent responder stage of com-
        petence development, 98,
        111–113
    decision-making simulations,
        112–113
    formulating plans, 111–112
    stress and, 113
Conditioning, physical, 125
Conflict
    control orientation as source of,
        35–40, 44
    in emergency services, 36–37
    home stress and, 64
Conflict management, 36–40
    negative methods of, 37–39
    positive methods of, 39–40
Content of work, sharing, 72–73
Control orientation
    conflict and, 35–40, 44
    nature of, 21–22
    negative aspects of, 30–31, 35–36
Conveying interest, 75–76
Counseling
    critical incident stress debriefing
        (CISD), 53–61, 138
    in stress management, 138

Cousins, Norman, 129, 130
Critical incidents
    effects of, 49–51
    nature of, 53–56
Critical incident stress debriefing
        (CISD), 53–61, 138
    accomplishment of, 53
    availability of, 59–61
    CISD team, 56
    nature of critical incidents, 53–56
    seven steps of, 56–58

de Becker, Gavin, 136
Decision-making simulations,
        112–113
Dedication, 27, 130
Diabetes, 6
*Diagnostic and Statistical Manual
        of Mental Disorders* (DSM), 48
Dialogue, in conflict management,
        39, 69–77. *See also*
        Communication skills
Diet
    sleep and, 92, 93
    stress levels and, 127–128
Difficult people, 30–31
Digestive disorders, 83
Divorce, 53, 82–83
Dreyfus Model of Skill Acquisition,
        97–98, 105–117
    advanced beginner stage, 98,
        107–111
    aspects of the situation and,
        107–108, 109
    competent responder stage, 98,
        111–113
    expert stage, 98, 115–117
    novice stage, 98, 105–107
    proficient responder stage, 98,
        113–115
Drug use, 53

Early warning signal, 115
Earplugs, 91
Emotional anesthesia, 49, 51
Employee assistance programs
        (EAPs), 138
Encouragement, 139
Endogenous circadian pacemaker
        (ECP), 84–86

Energy bars, 128
Epstein, Seymour, 15–16
Exercise
  benefits of regular, 92, 125–126,
    127
  sleep and, 92
Exhaustion phase, 2–4
Expectations. *See* High expectations
Experience, reaction to stress and,
  15–16, 50–51
Experiential learning, 119
Expert responder stage of competence development, 98,
  115–117
  advancement to, 119
  contribution of experts, 117
  intuitive responses, 115–117
Expressing confidence, 77

Family orientation, 27–28, 66–67.
  See also Home stress
Feedback, value of, 69–70
Fenz, Walter D., 15–16
Fight or flight response, 4–5
Financial issues, 137
Firehouse magazine, 20
Flashbacks, 49, 51
Flexibility, attentional styles and,
  103–105, 113
Fluids, stress levels and, 128–129
Food, 92, 93, 127–128
Frankl, Viktor E., 34, 139
Freud, Sigmund, 34

General adaptation syndrome
  (GAS), 2–4, 124
  long-term example of, 4
  phases of, 2–4
  short-term example of, 3–4
Guns, at home, 67

Headaches, 6
Heart disease, 6, 83
Hierarchy of needs (Maslow),
  122–124
High blood pressure, 6, 83
High expectations
  as cause of stress, 41–44
  impact on home life, 71, 77
  nature of, 29–30

negative aspects of, 31, 41–44
Hobbies, 130
Holmes, Thomas, 9–14
Holmes and Rahe Social
  Readjustment Rating Scale,
  9–14, 137
Home stress, 63–79
  addressing behavior problems
    and, 77–78
  communication as factor in,
    69–77
  job stress carried home and,
    68–69
  loyalty to job and, 66–67
  public opinion and, 69
  risk of danger and, 67
  understanding source of home
    life problems, 63–64
  work schedules and, 65–66, 93
Hostile aggression, 38–39
Humor, 130
Hydration, stress levels and,
  128–129
Hypertension, 6, 83
Hypnosis, 134

Immune system, 6
Inner voice, 136
Instincts, 115–117, 136
Intuitive responses, 115–117, 136
Irritable bowel syndrome (IBS), 6
Isolation, 31, 69

Jacobson, Edmund, 133
Job stress, 33–45
  carrying home, 68–69
  control orientation and conflict,
    35–40, 44
  costs of, 35
  high expectations and, 41–44, 77
  nature of, 33–34
  occupational stress versus, 33–34
  stress management and, 34,
    121–140
  symptoms of overstress, 34–35
John Wayne syndrome, 52, 61, 68
Journal writing, 130–131
Jung, Carl, 34

Kübler-Ross, Elisabeth, 34

"Larks," 89
Letting go, in conflict management, 40
Light, circadian rhythm and, 85–86, 91
Listening skills
    of emergency responders, 73–74
    of significant others, 77
Long sleepers, 84
Loyalty to job, home stress and, 66–67

Macho attitude, 52, 61, 68
Magnifying, 42
Martial arts, 133
Maslow, Abraham, 34, 122–124
Massage therapy, 133
Mastery, 112
Maxims, 113–114
Medications, 92–93
    alerting, 92
    sleeping, 93
Meditation, 132
Melatonin, 93
Menstrual irregularities, 83
Mental imagery, 133
Mentoring, 117–118
Migraine headaches, 6
Mitchell, Jeffrey T., 20, 53–54, 55
Money management, 137
Music therapy, 133

Naps, 90–91
Narrow external attentional style, 100–102, 105–107, 113–115
Narrow internal attentional style, 100, 101
National Institute of Mental Health (NIMH), 52–53
National Safety Council, 35
National Sleep Foundation, 83, 84, 90
Needs hierarchy (Maslow), 122–124
Negativity
    in the department, 43–44
    home stress and, 69
    negative self-talk, 42–43
    negative thinking, 31, 42, 44, 139
    public opinion and, 69
Nervousness, 49, 52
Nideffer, Robert, 99, 104, 134

Night shift workers, 82–83
Noise
    sleep hygiene and, 91
    stress levels and, 126–127
Non-rapid eye movement (NREM), 86
Novice stage of competence development, 98, 105–107
    role of rules in, 105–107, 118
    stress and, 107
Nutrition, 92, 93, 127–128

Occupational stress
    job stress versus, 33–34
    nature of, 33
On-call status, home stress and, 66
Open dialogue, 39
Overstress, symptoms of, 34–35
"Owls," 89

Paraphrasing, 73–74
Perfectionism, 29–30, 31, 41–42, 43
Persistent anxiety, 49, 52
Personality traits. *See* Traits of emergency responders
Persuasion, in conflict management, 39–40
Pessimism, 42
Physical stress, 2–7, 98–99, 124–129
    chronic stress, 5–6
    diet and, 92, 93, 127–128
    fight or flight response in, 4–5
    fluids and, 128–129
    general adaptation syndrome (GAS) and, 2–4, 124
    noise and, 91, 126–127
    physical activity demands and, 124–126
    role in illness and disease, 6, 82–83
    sleep and, 129
Plan formulation, 111–112
Positive attitude, 43, 139
Post-traumatic stress disorder (PTSD)
    categories of symptoms, 49, 51–52
    critical incident stress debriefing (CISD) and, 53–61
    nature of, 48–49
    responses to symptoms of, 52–53

Prejudgment, 73–74
Problem-solving approach, 78
Process debriefing, 53
Proficient responder stage of competence development, 98, 113–115
   case studies in, 114–115
   early warning signals and, 115
   maxims of, 113–114
   stress and, 115
Progressive Muscular Relaxation (PMR), 133–134
Psychological first aid, 59–60
Psychological stress, 7–8, 129–134
   diversions and distractions and, 130
   journal writing and, 130–131
   power of the mind and, 129
   relaxation techniques and, 132–134
   rest and, 129–130
   spirituality and, 132
Public opinion, home stress and, 69

Rahe, Richard, 9–14
Raphael model, 53
Rapid eye movement (REM), 86
Reaction, stress as, 2
Re-experiencing events, 49, 51
Relaxation techniques, 132–134
Reliving events, 49, 51
Rescue orientation, 30
Resistance phase, 2–4
Rest
   naps and, 90–91
   stress levels and, 129–130
Risk-taking behavior, 15–16, 26–27, 44, 67, 71
Role identification, 24
Routines, in sleep hygiene, 89–90
Rules, in novice stage of competence development, 105–107, 118

Seaward, Brian Luke, 6
Self-comparison, 42
Self-esteem, 31, 42, 64, 77, 130
Self-hypnosis, 134
Selye, Hans, 1–2, 4
Sensational activities, 15–16, 44

Sharing content, in home life, 72–73
Sheehy, Gail, 34, 137
Shift work
   home stress and, 65–66, 93
   night shift workers, 82–83
   sleep deprivation and, 81–84, 87, 88
Short sleepers, 84
"Should"-ing, 42
Skin rashes, 6
Sky divers, 15–16, 44
Sleep, 81–94
   requirements for, 83–84
   shift work and sleep deprivation, 81–84, 87, 88
   sleep hygiene and, 89–93
   sleep-wake cycle, 84–89
   stress levels and, 129–130
Sleep apnea, 87
Sleep debt, 84
Sleep drive buildup, 86–87
Sleep hygiene, 89–93
   diet in, 92, 93
   environmental factors in, 91
   establishing routines in, 89–90
   exercise in, 92
   naps and, 90–91
   sleep-related medications, 92
Sleep inertia, 90
Sleep problems, 49, 52
Sleep-related medications, 92–93
   alerting medications, 92
   sleeping medications, 93
Slow-motion effect, 7–8
Social engineering, 137
Social Readjustment Rating Scale Self-Test, 13
Span of control, 40
Spending time, in home life, 74–75
Spirituality, 132
Stomach disorders, 83
Stress. *See also* Acute stress disorder (ASD); Post-traumatic stress disorder (PTSD)
   choice and reaction to, 14–15, 50, 129
   competence level and, 102–105, 107, 109–110, 113, 115
   control orientation and, 30–31, 35–40

defined, 1–2, 10
different responses to same
    event, 14–16
experience and reaction to,
    15–16, 50–51
high expectations and, 31, 41–44
at home. *See* Home stress
from mismatch between skills
    and demands of situation, 98,
    102–105
physical, 2–7, 98–99, 124–129
from positive and negative
    events, 9–14
psychological, 7–8, 129–134
as reaction, 2
Stress management, 121–140
    interaction with others and,
        135–139
    Maslow's hierarchy of needs and,
        122–124
    for physical stress, 124–129
    for psychological stress, 129–134
    types of programs, 34, 121–122
Stroke, 6
Suicide, 53, 54
Surrender, in conflict management,
    38
Sympathetic nervous system, 6

Temperature, sleep hygiene and, 91
Temporomandibular joint dysfunc-
    tion (TMJ), 6
Tension headaches, 6
Terrorist attacks of September 11,
    2001, 67
Thrill seekers, 15–16, 26–27, 44, 71
Time management, 136–137
Toxic thoughts, 42–43
Traditionalism, 22–23
Training approaches
    case studies and, 114–115
    experiential learning, 119
    mentoring, 117–118
    rules and, 105–107, 118
    simulations, 112
    traditional curricula, 118–119
Traits of emergency responders,
    19–32

action orientation, 25–26, 71,
    130
adrenaline rush and, 19–21, 25,
    32, 64, 71
attention to detail, 8, 22
control orientation, 21–22,
    30–31, 35–40, 44
dedication, 27, 130
family orientation, 27–28, 66–67
high expectations, 29–30, 31,
    41–44, 71, 77
impact on home life, 71
negative aspects of, 30–31
positive aspects of, 44
rescue orientation, 30
risk-taking behavior, 15–16,
    26–27, 44, 67, 71
strong role identification, 24
traditionalism, 22–23
Troubleshooting, analytical,
    116–117
Tunnel vision, 8, 22

U.S. National Institute of Mental
    Health, 59
U.S. Navy Seals, 99
Victimization, 42
Vision, of experts, 117
Visualization, 133
Vitamin supplements, 128

Water intake, stress levels and,
    128–129
Weapons, at home, 67
"White noise" generators, 91
Withdrawal, in conflict manage-
    ment, 37
Work-related accident fatality rate,
    67
Work schedules
    home stress and, 65–66, 93
    sleep deprivation and, 81–84, 87,
        88
World Health Organization (WHO),
    59

Yerkes-Dodson curve, 7
Yoga, 132